Student Solutions Manual

Physics for Scientists and Engineers, 2nd Edition

Hawkes/Iqbal/Mansour/Milner-Bolotin/Williams

Custom Edition for University of British Columbia

NELSON

NELSON

COPYRIGHT © 2020 by Nelson Education Ltd.

Printed and bound in Canada
1 2 3 4   22 21 20 19

For more information contact Nelson Education Ltd., 1120 Birchmount Road, Toronto, Ontario, M1K 5G4. Or you can visit our Internet site at nelson.com

ALL RIGHTS RESERVED. No part of this work covered by the copyright herein may be reproduced, transcribed, or used in any form or by any means—graphic, electronic, or mechanical, including photocopying, recording, taping, Web distribution, or information storage and retrieval systems—without the written permission of the publisher.

For permission to use material from this text or product, submit all requests online at cengage.com/permissions. Further questions about permissions can be emailed to permissionrequest@cengage.com

Every effort has been made to trace ownership of all copyrighted material and to secure permission from copyright holders. In the event of any question arising as to the use of any material, we will be pleased to make the necessary corrections in future printings.

This textbook is a Nelson custom publication. Because your instructor has chosen to produce a custom publication, you pay only for material that you will use in your course.

ISBN-13: 978-0-17-678603-8
ISBN-10: 0-17-678603-1

Consists of Selections from:

Student Solutions Manual for Hawkes' Physics for Scientists and Engineers
Robert Hawkes, Javed Iqbal, Firas Mansour, Marina Milner-Bolotin, Peter Williams
ISBN-10: 0-17-677046-1, © 2018

Cover Credit:

R.T. Wohlstadter/Shutterstock

Contents

Chapter 4: Motion in Two and Three Dimensions ... 1

Chapter 12: Fluids ... 17

Chapter 13: Oscillations ... 34

Chapter 14: Waves ... 55

Chapter 15: Interference and Sound ... 73

Chapter 29: Physical Optics ... 84

Chapter —MOTION IN TWO AND THREE DIMENSIONS

1. (a) D

   (b) B

   (c) C

   (d) A

3. c

5. b

7. b. When the boat is pointed straight across the river,

   $$\theta = \tan^{-1}\left(\frac{12}{60}\right) = 11.31°$$

   Since $\tan\theta = \dfrac{v_{river}}{v_{boat}}$,

   $$v_{boat} = 5v_{river}$$

   For the boat to have total velocity directly toward the target, its launch angle relative to the target should be

   $$\theta' = \sin^{-1}\left(\frac{v_{river}}{v_{boat}}\right) = 11.54°$$

   $$\theta' > \theta$$

   Thus, the boat arrives north of the target.

9. d

11. d. Since both marbles are dropped from the same height and the initial vertical velocity of the rolled marble is zero, the vertical distance travelled by the two marbles during the time the first marble moves horizontally for 1 m is the same.

13. d. $d = 2.25 \times 2\pi \times \dfrac{4}{\pi}$ m $\neq 4$ m

    $$\Delta r = \sqrt{\left(\frac{4}{\pi}\,\text{m}\right)^2 + \left(\frac{4}{\pi}\,\text{m}\right)^2} \neq \frac{2\sqrt{2}}{\pi}\,\text{m}$$

15. d. In the apple's frame of reference, the relative acceleration of the bullet is zero. Thus, if the bullet was initially heading toward the apple, it will always be heading toward the apple.

17. a. The vertical distance covered by the projectile fired upward is

$$h = v\sin(30°)t - \frac{1}{2}gt^2$$

The vertical distance by which the second projectile will drop is

$$h' = -\frac{1}{2}gt^2$$

Hence, the vertical distance between the two projectiles is
$$h - h' = v\sin(30°)t$$

The horizontal distance covered by the first projectile is

$$x = v\cos(30°)t$$

The distance covered by the second projectile is

$$x' = vt$$

Hence, the horizontal distance between the two projectiles is

$$x' - x = vt[1 - \cos(30°)]$$

Thus, the horizontal distance between the two will increase with time.

19. a (since the ball will be travelling with the same horizontal speed)

21. Both will hit the ground at the same time. Both objects have the same (zero) initial vertical component of velocity, so they take the same amount of time to travel the same vertical distance.

23. Tangential acceleration: tangent to the bowl and pointing in the direction of motion; radial acceleration: pointing toward the centre of the bowl

25. (a) We are given $v_1 = 112$ km/h = 31.1 m/s. Thus,

$$x_1 = v_1 t_1 = 31.1 \text{ m/s} \times 3.00 \text{ s} = 93.3 \text{ m}$$
$$x_2 = v_2 t_2 = 78.0 \text{ m/s} \times 2.00 \text{ s} = 156 \text{ m}$$

We are also given that $x_3 = 11.0$ m. Thus,

$$x = x_1 + x_2 + x_3 = 93.3 \text{ m} + 156 \text{ m} + 11.0 \text{ m} = 260.3 \text{ m}$$

For the deceleration phase we have

$$a = \frac{v_f^2 - v_2^2}{2x_3} = \frac{0 - (78.0 \text{ m/s})^2}{2 \times 11.0 \text{ m}} = -276.5 \text{ m/s}^2$$

Thus,

$$t_3 = \frac{v_f - v_2}{a} = \frac{0 - 78 \text{ m/s}}{-276.5 \text{ m/s}^2} = 0.282 \text{ s}$$

$$t = t_1 + t_2 + t_3 = 3.00 \text{ s} + 2.00 \text{ s} + 0.282 \text{ s} = 5.28 \text{ s}$$

$$v = \frac{x}{t} = \frac{260.3 \text{ m}}{5.28 \text{ s}} = 49.3 \text{ m/s}$$

(b) To calculate the displacement, we use the cosine law:

$$d = \sqrt{x_1^2 + (x_2 + x_3)^2 - 2x_1(x_2 + x_3)\cos(138°)}$$
$$= \sqrt{(93.3 \text{ m})^2 + (156 \text{ m} + 11.0 \text{ m})^2 - 2(93.3 \text{ m})(156 \text{ m} + 11.0 \text{ m})\cos(138°)}$$
$$= 244.4 \text{ m}$$

$$v_{avg} = \frac{d}{t} = \frac{244.4 \text{ m}}{5.28 \text{ s}} = 46.3 \text{ m/s}$$

To calculate the angle of displacement, we use the sine law:

$$\frac{\sin \theta}{x_2 + x_3} = \frac{\sin(138°)}{d}$$

$$\theta = 27.2°$$

$$\vec{v}_{avg} = 46.3 \text{ m/s } [27.2° \text{ from initial direction}]$$

(c) $\vec{a}_{avg} = \frac{\vec{v}_f - \vec{v}_i}{t} = \frac{0 - 31.1 \text{ m/s}}{5.28 \text{ s}} = 5.89 \text{ m/s}^2$ [antiparallel to original velocity]

(d) This was calculated in part (a): $t_3 = 0.282$ s.

In many of the solutions below, units are omitted until the final answer to make calculations simpler.

27. The trajectory of the particle is given below. We are given that

$$x = (8.00 \text{ m} \cdot \text{s}^{-1})t, \quad y = \frac{1}{(16.0 \text{ m}^{-1} \cdot \text{s}^{-1})t}$$

Thus,

$$v(t) = \sqrt{8.00^2 + \left(-\frac{1}{16.0 t^2}\right)^2} = \sqrt{64.0 + \frac{1}{256 t^4}}$$

$$v(3.00) = 8.00 \text{ m/s}$$

To calculate the acceleration, we differentiate the velocity:

$$a_x = 0, \quad a_y = \frac{1}{8.00t^3}$$

$$a(t) = \sqrt{0^2 + \left(\frac{1}{8.00t^3}\right)^2} = \frac{1}{8.00t^3}$$

$$a(3.00) = 0.004\ 63\ \text{m/s}^2$$

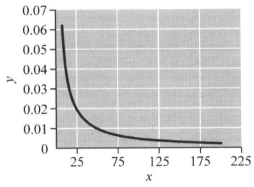

29. The projectile takes 1.25 s to reach half of the total horizontal distance as well as the maximum height. For the horizontal distance, we have

$$v_{x_0} = v_{\text{launch}} \cos\theta = \frac{2.15\ \text{m}}{1.25\ \text{s}} = 1.72\ \text{m/s} \quad (1)$$

For the vertical distance, we have

$$v_{y_0} = v_{\text{launch}} \sin\theta = 9.81\ \text{m/s}^2 \times 1.25\ \text{s} = 12.3\ \text{m/s} \quad (2)$$

Dividing (2) by (1), we get

$$\tan\theta = \frac{12.3\ \text{m/s}}{1.72\ \text{m/s}}$$

$$\theta = 82.0° \quad \text{(angle of launch)}$$

Thus,

$$v_{\text{launch}} = \frac{12.3\ \text{m/s}}{\sin(82.0°)} = 12.4\ \text{m/s}$$

The minimum speed occurs at the apex, where the vertical speed goes to zero and the total speed is 1.72 m/s:

$$\vec{a}_{\text{avg}} = 9.81\ \text{m/s}^2\ [\text{down}]$$

$$\vec{v}_{\text{avg}} = \frac{\Delta\vec{x}}{t_{\text{total}}} = \frac{4.30\ \text{m}}{2.5\ \text{s}} = 1.72\ \text{m/s in the horizontal direction}$$

31. The position of the object in the x-direction is

$$x = (v\cos\theta)t$$

$$t = \frac{x}{v\cos\theta} \qquad (1)$$

The position of the object in the y-direction is

$$y = vt\sin\theta + \frac{1}{2}gt^2$$

Substituting the value of t in this equation from (1), we get

$$y = v\left(\frac{x}{v\cos\theta}\right)\sin\theta + \frac{1}{2}g\left(\frac{x}{v\cos\theta}\right)^2$$

$$y = x\tan\theta + \frac{g}{2v^2\cos^2\theta}x^2$$

Since this is the equation of a parabola, the object's trajectory is parabolic.

33. (a) $\theta = \dfrac{d}{R_{Earth}} = \dfrac{9500 \text{ km}}{6371 \text{ km}} = 1.49 \text{ rad} = 85°$

Earth's rotation rate about its axis is

$\omega = 360°/\text{day}$

Thus,

$$t = \frac{\theta}{\omega} = \frac{85°}{360°/\text{day}} = 0.24 \text{ days} = 5.8 \text{ h}$$

(b) We are given $v = 970$ km/h $= 269$ m/s:

$$a_r = \frac{v^2}{r} = \frac{(269 \text{ m/s})^2}{6\,371\,000 \text{ m} + 10\,000 \text{ m}} = 0.011 \text{ m/s}^2$$

35. At the equator, the Sun's rotation period is $T = 24.47$ days. The radius of the Sun is $R_{Sun} = 6.955 \times 10^8$ m. The speed of a particle moving on the Sun's equator is

$$v = \frac{2\pi R_{Sun}}{T} = \frac{2\pi \times 6.955 \times 10^8 \text{ m}}{24.47 \times 24 \times 3600 \text{ s}} = 2067 \text{ m/s}$$

Thus, the acceleration is

$$a_r = \frac{v^2}{R_{Sun}} = \frac{(2067 \text{ m/s})^2}{6.955 \times 10^8 \text{ m}} = 6.143 \times 10^{-3} \text{ m/s}^2$$

37. $a_r = \dfrac{v^2}{r}$

$v = \sqrt{a_r r} = \sqrt{2\times 10^6 \times 9.81 \text{ m/s}^2 \times 0.0450 \text{ m}} = 940. \text{ m/s}$

The speed divided by the circumference gives the speed of rotation:

rotation speed $= \dfrac{v}{2\pi r} = \dfrac{940. \text{ m/s}}{2\pi \times 0.0450 \text{ m}} = 3300$ rotations/s

39. We are given $\theta = 180° - 30° = \dfrac{5\pi}{6}$ rad. The distance covered along the perimeter is

$d = \theta r = \dfrac{5\pi}{6}$ rad $\times 4.0$ m $= 10.5$ m

Thus,

$a_t = \dfrac{v_f^2 - v_i^2}{2d} = \dfrac{(8.00 \text{ m/s})^2 - (12.0 \text{ m/s})^2}{2\times 10.5 \text{ m}} = -3.81 \text{ m/s}^2$

$a_r = \dfrac{v^2}{r} = \dfrac{(8.00 \text{ m/s})^2}{4.0 \text{ m}} = 16.0 \text{ m/s}^2$

$a = \sqrt{a_t^2 + a_r^2} = \sqrt{(3.81 \text{ m/s}^2)^2 + (16.0 \text{ m/s}^2)^2} = 16.4 \text{ m/s}^2$

$\theta = \tan^{-1}\left(\dfrac{3.81}{16.0}\right) = 13.4°$ with respect to the radial direction

41. The tangential acceleration is given by

$a_t = \dfrac{v_f^2 - v_i^2}{2S} = \dfrac{(4.0 \text{ m/s})^2 - 0}{2(3\times 2\pi \times 0.72 \text{ m})} = 0.59 \text{ m/s}^2$

The required condition is $a_r = a_t$. Thus,

$\dfrac{v_f^2}{r} = 0.59 \text{ m/s}^2$

$v_f = 0.65$ m/s

The time to achieve this velocity is

$t = \dfrac{v_f}{a_t} = \dfrac{0.65 \text{ m/s}}{0.59 \text{ m/s}^2} = 1.1$ s

Thus,

$x = \dfrac{v_f^2 - v_i^2}{2a_t} = \dfrac{(0.65 \text{ m/s})^2 - 0}{2\times 0.59 \text{ m/s}^2} = 0.36$ m

This is equivalent to $\dfrac{0.36 \text{ m}}{2\pi(0.72 \text{ m})} = 0.079$ turns.

43. Since the horizontal speed of the acrobat is equal to the speed of the car, she will land in the parade car. We are given that $v_x = 10.0$ km/h $= 2.78$ m/s. The time to land back in the parade car can be found from

$$\Delta y = v_{y_0} t - \frac{1}{2} g t^2 = 0$$

$$t = 2 \times \frac{14 \text{ m/s}}{9.81 \text{ m/s}^2} = 2.85 \text{ s}$$

Thus,

$$x = v_x t = 2.78 \text{ m/s} \times 2.85 \text{ s} = 7.9 \text{ m}$$

45. The speed of the swimmer relative to the ground during each interval is

$$v_{s,g,1} = v_{c,g} + v_{s,c}$$

$$v_{s,g,2} = v_{c,g} - v_{s,c}$$

The displacement of the swimmer relative to the ground after 10.0 min is

$$x_{s,g} = v_{s,g,1} \times 3.0 \times 10^2 \text{ s} + v_{s,g,2} \times 3.0 \times 10^2 \text{ s} = 500. \text{ m}$$

Thus,

$$(v_{c,g} + v_{s,c}) \times 3.0 \times 10^2 \text{ s} + (v_{c,g} - v_{s,c}) \times 3.0 \times 10^2 \text{ s} = 500. \text{ m}$$

$$v_{c,g} = \frac{500. \text{ m}}{2 \times 3.0 \times 10^2 \text{ s}} = 0.83 \text{ m/s}$$

47. Describe the motion of the arrow relative to the watermelon. The initial velocity of the arrow relative to the melon was such that it would pass a point 5.0 m above the melon after travelling a horizontal distance of 90. m. Because the acceleration of the arrow relative to the melon is zero, the melon will see the arrow travel in a perfectly straight line and pass 5.0 m above it.

49. Initially, the velocity of the teddy bear with respect to the ground has components:

$$v_x = 6.0 \text{ m/s} \times \cos(37°) + 2.7 \text{ m/s} = 7.5 \text{ m/s}$$

$$v_y = 6.0 \text{ m/s} \times \sin(37°) = 3.6 \text{ m/s}$$

$$v = \sqrt{v_x^2 + v_y^2} = \sqrt{(7.5 \text{ m/s})^2 + (3.6 \text{ m/s})^2} = 8.3 \text{ m/s}$$

At the time the teddy bear hits the ground, its horizontal velocity is unchanged, and its vertical velocity is

$$v_{y_f} = \sqrt{v_{y_f}^2 + 2\vec{a} \cdot \Delta \vec{y}} = \sqrt{(3.6 \text{ m/s})^2 + 2 \times 9.81 \text{ m/s}^2 \times 3.0 \text{ m}} = 8.5 \text{ m/s}$$

So, when the bear hits the ground, its speed is

$$v = \sqrt{v_x^2 + v_y^2} = \sqrt{(7.5 \text{ m/s})^2 + (8.5 \text{ m/s})^2} = 11 \text{ m/s}$$

51. We are given that $v_e = 34$ km/h $= 9.44$ m/s.

    The distance travelled by the eagle in 1.0 s is 9.44 m. Switching to the frame of reference of the eagle, a person is throwing a stone and trying to hit a point 23 m up and 9.94 m (9.44 m + 50 cm) over. The launch velocity in the eagle's frame of reference is

    $v_{x_0} = v\cos(38°) - 9.44$ m/s

    $v_{y_0} = v\sin(38°)$

    For the stone to hit the point 50. cm in front of the eagle,

    $v_{x_0} t = 9.94$ m $= (v\cos(38°) - 9.44 \text{ m/s})t$ \hfill (1)

    $v_{y_0} t - \frac{1}{2}gt^2 = 23$ m $= v\sin(38°)t - \frac{1}{2}gt^2$ \hfill (2)

    One can use a graphing calculator to find that there is no solution to this problem. This means that there is no speed that can give the necessary altitude and horizontal displacement at the same time for a given angle.

53. The time taken by the papaya to hit the ground is

    $t = \sqrt{\dfrac{2 \times 14.0 \text{ m}}{9.81 \text{ m/s}^2}} = 1.69$ s

    Therefore, the stone hits the papaya after 1.69 s − 0.900 s = 0.79 s. To hit the papaya, which dropped from rest, aim directly at it assuming zero acceleration. Because there is zero relative acceleration between the stone and the papaya, they will collide. For the monkey that is 11.0 m from the tree,

    $v_{x_i} = \dfrac{11.0 \text{ m}}{0.79 \text{ s}} = 13.9$ m/s

    $v_{y_i} = \dfrac{14.0 \text{ m}}{0.79 \text{ s}} = 17.7$ m/s

    When the stone hits the papaya, its vertical speed and total speed are

    $v_y = 17.7 \text{ m/s} - 9.81 \text{ m/s}^2 \times 0.79 \text{ s} = 10.0$ m/s

    $v = \sqrt{(13.9 \text{ m/s})^2 + (10.0 \text{ m/s})^2} = 17.1$ m/s

    Similarly, the monkey throwing from 16 m throws so the speed at collision is 22.6 m/s.

55. (a) We are given that $y = 60.0t^2 - 7.00t^3 + 4.00t + 120.$, omitting units for now. For the vertical velocity, we differentiate this with respect to time:

    $v_y = 120.t - 21.0t^2 + 4.00$

At the maximum height, the vertical velocity is zero; that is,

$$120.t - 21.0t^2 + 4.00 = 0$$
$$t = 5.75 \text{ s}$$

Thus,

$$y_{max} = 60.0(5.75)^2 - 7.00(5.75)^3 + 4.00(5.75) + 120. = 796 \text{ m}$$

(b) The time taken by the rocket to travel 120. m below the point of launch can be calculated from

$$y = 60.0t^2 - 7.00t^3 + 4.00t + 120. = 0$$
$$t = 8.85 \text{ s}$$

The horizontal speed is a constant 3 m/s, and the vertical speed is

$$v_y(8.85 \text{ s}) = 120.(8.85) - 21.0(8.85)^2 + 4.00 = -579 \text{ m/s}$$

Thus,

$$v_{total} = \sqrt{(3.00 \text{ m/s})^2 + (-579 \text{ m/s})^2} = 579 \text{ m/s}$$

At $t = 8.85$ s, the horizontal displacement is

$$x(8.85 \text{ s}) = 1.40 \times 10^3 \text{ m} + (3.00 \text{ m/s})(8.85 \text{ s}) = 1430 \text{ m}$$

(c) We see that the acceleration in the x-direction is zero. Therefore, the total acceleration is the acceleration in the y-direction, which can be obtained by differentiating the expression for velocity in the y-direction:

$$a_y = 120. - 42.0t$$
$$a(0) = 120. \text{ m/s}^2$$
$$a(8.85 \text{ s}) = -252 \text{ m/s}^2$$

Since the acceleration varies linearly with time, it will have its maximum (absolute) value right before it hits the ground.

57. The component of velocity in the xy-plane is

$$v_{xy} = 14.0 \text{ m/s} \times \cos(90.0° - 29.0°) = 14.0 \text{ m/s} \times \cos(61.0°)$$

The x-component is then given by

$$v_x = 14.0 \text{ m/s} \times \cos(61.0°)\cos(40.0°) = 5.20 \text{ m/s}$$

Since there is no acceleration in the x-direction, the x-component is

$$x = 5.20 \text{ m/s} \times t$$

Similarly, for the y-direction we have

$v_y = 14.0 \text{ m/s} \times \cos(61.0°)\sin(40.0°) = 4.36 \text{ m/s}$

Thus,

$y = 4.36 \text{ m/s} \times t$

The initial velocity in the z-direction is

$v_z = 14.0 \text{ m/s} \times \sin(61.0°) = 12.2 \text{ m/s}$

Since there is acceleration in the z-direction, we have

$z = (12.2 \text{ m/s}) \times t - \frac{1}{2}(9.81 \text{ m/s}^2) \times t^2$

59. (a) The maximum height reached above the tree house is

$h_1 = \frac{v_y^2}{2g} = \frac{((35.0 \text{ m/s})\sin(37.0°))^2}{2 \times 9.81 \text{ m/s}^2} = 22.6 \text{ m}$

Hence, the maximum height from the ground is $h_{max} = 4.0 \text{ m} + 22.6 \text{ m} = 26.6 \text{ m}$.

(b) The time needed to reach the ground can be found using the quadratic formula:

$y(t) = -\frac{1}{2}(9.81 \text{ m/s}^2) \times t^2 + (35.0 \text{ m/s}) \times \sin(37.0°)t + 4.00 = 0$

$t = 4.48 \text{ s}$

The horizontal distance covered is

$x(4.48 \text{ s}) = 35.0 \text{ m/s} \times \cos(37.0°)(4.48 \text{ s}) = 125 \text{ m}$

(c) $v_f = \sqrt{v_i^2 + 2\vec{a} \cdot \Delta \vec{x}} = \sqrt{(35.0 \text{ m/s})^2 + 2 \times 9.81 \text{ m/s}^2 \times 4.00 \text{ m}} = 36.1 \text{ m/s}$

61. $x = v_0 t \cos\theta$ (1)

$y = v_0 t \sin\theta + \frac{1}{2}at^2$ (2)

Solving (1) for t and substituting into (2) gives

$y = x\tan\theta + \frac{ax^2}{2v_0^2 \cos^2\theta}$

Since at maximum range, $y = 0$, the above equation becomes

$$0 = x\tan\theta - \frac{gx^2}{2v_0^2 \cos^2\theta}$$

$$x\left(\tan\theta - \frac{gx}{2v_0^2 \cos^2\theta}\right) = 0$$

$$x = \frac{2v_0^2 \sin\theta\cos\theta}{g} = \frac{v_0^2 \sin(2\theta)}{g}$$

The maximum range occurs when $\frac{dx}{d\theta} = 0$. Thus,

$$\frac{2v_0^2 \cos 2\theta}{g} = 0$$

$$\theta = 45°$$

The apex of flight occurs when x is half its maximum range:

$$x_{y\text{-max}} = \frac{1}{2}x_{\max} = \frac{1}{2}\frac{v_0^2 \sin(2\times 45°)}{g} = \frac{v_0^2}{2g}$$

$$y_{\max} = \frac{v_0^2 \sin^2\theta}{2g} = \frac{v_0^2 \sin^2(45°)}{2g} = \frac{v_0^2}{4g}$$

63. (a) We are given that $\omega = 10.\text{ rev/min} = 20\pi\text{ rad}/60\text{ s} = \pi/3\text{ rad/s}$.

Imagine the first wheel centred at $(-0.32\text{ m}, 0)$; the position is

$$\vec{r}_1 = 0.32\text{ m}\times\left[\cos\left(\frac{\pi}{3\text{ s}}t\right)\hat{i} + \sin\left(\frac{\pi}{3\text{ s}}t\right)\hat{j}\right] - (0.32\text{ m})\hat{i}$$

The position of the other point is reflected in the y-axis ($\hat{i} \to -\hat{i}$):

$$\vec{r}_2 = 0.32\text{ m}\times\left[-\cos\left(\frac{\pi}{3\text{ s}}t\right)\hat{i} + \sin\left(\frac{\pi}{3\text{ s}}t\right)\hat{j}\right] + (0.32\text{ m})\hat{i}$$

$$\vec{r}_{\text{rel}} = \vec{r}_2 - \vec{r}_1 = 0.32\text{ m}\left(2 - 2\cos\left(\frac{\pi}{3\text{ s}}t\right)\right)\hat{i}$$

(b) The relative velocity is the derivative of the relative position:

$$\vec{v}_{\text{rel}} = 0.32\text{ m}\times\frac{2\pi}{3\text{ s}}\sin\left(\frac{\pi}{3\text{ s}}t\right)\hat{i} = 0.67\text{m/s}\times\sin\left(\frac{\pi}{3\text{ s}}t\right)\hat{i}$$

(c) The relative acceleration is the derivative of the relative velocity:

$$\vec{a}_{rel} = 0.70 \text{ m/s}^2 \times \cos\left(\frac{\pi}{3 \text{ s}}t\right)\hat{i}$$

65. The velocities of the Boeing 747 and the smaller plane are

$$\vec{v}_B = 790 \text{ km/h}(-\cos(11°)\hat{i} + \sin(11°)\hat{j}) = (-775 \text{ km/h})\hat{i} + (151 \text{ km/h})\hat{j}$$
$$\vec{v}_S = 430 \text{ km/h}(-\sin(23°)\hat{i} - \cos(23°)\hat{j}) = (-168 \text{ km/h})\hat{i} - (396 \text{ km/h})\hat{j}$$

The relative velocity of the small plane to the Boeing 747 is

$$\vec{v}_{rel} = \vec{v}_S - \vec{v}_B = (607 \text{ km/h})\hat{i} - (547 \text{ km/h})\hat{j}$$

If we say the small plane is at the origin, the position of the Boeing is

$$\vec{x}_B = -12 \text{ km}(\cos(17°)\hat{i} + \sin(17°)\hat{j}) = -(11.5 \text{ km})\hat{i} - (3.51 \text{ km})\hat{j}$$

Our task reduces to finding the distance between a line and a point. The line describing the path of the small plane in the Boeing's frame of reference is $547x + 607y = 0$, forgoing units. The shortest distance from a point (x_B, y_B) and a line $ax + by = 0$ is

$$d = \frac{|ax_B + by_B|}{\sqrt{a^2 + b^2}} = \frac{|547(-11.5) + 607(-3.51)|}{\sqrt{547^2 + 607^2}} = 10.3 \text{ km}$$

The point along the line closest to the Boeing has coordinates

$$x_{S_{closest}} = \frac{b^2 x_B - aby_B}{a^2 + b^2} = \frac{607^2(-11.5) - 547(607)(-3.51)}{547^2 + 607^2} = -4.60 \text{ km}$$

$$y_{S_{closest}} = \frac{-abx_B + a^2 y_B}{a^2 + b^2} = \frac{-547(607)(-11.5) + 547^2(-3.51)}{547^2 + 607^2} = 4.15 \text{ km}$$

This point is behind the original position of the small plane, so the closest approach happened in the past.

For the small plane to travel that distance in the Boeing's frame of reference,

$$t = \frac{\Delta x}{v_{rel}} = \frac{\sqrt{(-4.60 \text{ km})^2 + (4.15 \text{ km})^2}}{\sqrt{(607 \text{ km/h})^2 + (-547 \text{ km/h})^2}} = 7.59 \times 10^{-3} \text{ h} = 27.3 \text{ s}$$

27.3 s before $t = 0$, the planes passed within 10.3 km of each other.

67. (a) $a_r = \dfrac{v_1^2}{r} = 6g$

$\Rightarrow v_1 = \sqrt{6gr} = \sqrt{6 \times 9.81 \text{ m/s}^2 \times 20.0 \text{ m}} = 34.31$ m/s (velocity at the end of 5 turns)

$a_t = \dfrac{v_1^2 - v_0^2}{2S} = \dfrac{(34.29 \text{ m/s})^2 - 0}{2 \times 5 \times 2\pi \times 20.0 \text{ m}} = 0.9368$ m/s^2

$\Rightarrow t = \dfrac{v_1 - v_0}{a_t} = \dfrac{34.31 \text{ m/s} - 0}{0.9368 \text{ m/s}^2} = 36.6$ s

(b) $a_t = \dfrac{v_f - v_i}{t} = \dfrac{0 - 34.31 \text{ m/s}}{6.00 \text{ s}} = -5.72$ m/s^2

The velocity 3.00 s after the pod begins to decelerate is

$v_2 = v_1 + at = 34.31 \text{ m/s} - 5.72 \text{ m/s}^2 \times 3.00 \text{ s} = 17.15$ m/s

$\Rightarrow a_r = \dfrac{v_2^2}{r} = 14.71$ m/s^2

$\Rightarrow a = \sqrt{(5.72 \text{ m/s}^2)^2 + (14.71 \text{ m/s}^2)^2} = 15.8$ m/s^2

$\theta = \tan^{-1}\left(\dfrac{5.72}{14.71}\right) = 21.2°$ [away from radial direction]

(c) The distance covered during the constant-velocity phase is given by

$d_2 = 34.31 \text{ m/s} \times 2.00 \text{ s} = 68.62$ m

The distance covered during the deceleration phase is given by

$d_3 = \dfrac{v_f^2 - v_i^2}{2a_t} = \dfrac{0 - (34.31 \text{ m/s})^2}{-2 \times 5.72 \text{ m/s}^2} = 102.9$ m

$\Rightarrow N = \dfrac{68.62 \text{ m} + 102.9 \text{ m}}{2\pi \times 20.0 \text{ m}} = 1.36$ turns

$\Rightarrow N_{total} = 5 + 1.36 = 6.36$ turns

(d) $x(3.00 \text{ s}) = \dfrac{1}{2}(0.9368 \text{ m/s}^2)(3.00 \text{ s})^2 = 4.22$ m

$x(6.00 \text{ s}) = \dfrac{1}{2}(0.9368 \text{ m/s}^2)(6.00 \text{ s})^2 = 16.9$ m

$\theta(3.00 \text{ s}) = \dfrac{4.22 \text{ m}}{20.0 \text{ m}} = 0.211$ rad

$\theta(6.00 \text{ s}) = \dfrac{16.7 \text{ m}}{20.0 \text{ m}} = 0.834$ rad

$v(3.00 \text{ s}) = 0.9368 \text{ m/s}^2 \times 3.00 \text{ s} = 2.81 \text{ m/s}$

$v(6.00 \text{ s}) = 0.9368 \text{ m/s}^2 \times 6.00 \text{ s} = 5.62 \text{ m/s}$

$$\vec{a}_{avg} = \frac{\vec{v}_2 - \vec{v}_1}{t}$$

$$= \frac{(5.62 \text{ m/s})(\cos(0.845)\hat{i} + \sin(0.845)\hat{j}) - (2.81 \text{ m/s})(\cos(0.211)\hat{i} + \sin(0.211)\hat{j})}{3.00 \text{ s}}$$

$$= (0.327 \text{ m/s}^2)\hat{i} + (1.21 \text{ m/s}^2)\hat{j}$$

The magnitude of the average acceleration is

$$a_{avg} = \sqrt{(0.327 \text{ m/s}^2)^2 + (1.21 \text{ m/s}^2)^2} = 1.25 \text{ m/s}^2$$

(e) $x(4.00 \text{ s}) = \frac{1}{2}(0.9368 \text{ m/s}^2)(4.00 \text{ s})^2 = 7.49 \text{ m}$

$\theta(4.00 \text{ s}) = \frac{7.49 \text{ m}}{20.0 \text{ m}} = 0.374 \text{ rad}$

$$\vec{v}_{avg} = \frac{\vec{x}_2 - \vec{x}_1}{t}$$

$$= \frac{(20.0 \text{ m})(\cos(0.845)\hat{i} + \sin(0.845)\hat{j}) - (20.0 \text{ m})(\cos(0.374)\hat{i} + \sin(0.374)\hat{j})}{2.00 \text{ s}}$$

$$\vec{v}_{avg} = (-2.67 \text{ m/s})\hat{i} + (3.83 \text{ m/s})\hat{j}$$

The magnitude of the average velocity is

$$v_{avg} = \sqrt{(-2.67 \text{ m/s})^2 + (3.83 \text{ m/s})^2} = 4.67 \text{ m/s}$$

(f) $x_1 = \frac{0.9368 \text{ m/s}^2 \times (2.00 \text{ s})^2}{2} = 1.87 \text{ m}$

$x_2 = \frac{0.9368 \text{ m/s}^2 \times (5.00 \text{ s})^2}{2} = 11.71 \text{ m}$

$\Delta x = 11.71 - 1.87 = 9.84 \text{ m}$

This corresponds to an angle of

$\theta = \frac{9.84 \text{ m}}{20.0 \text{ m}} = 0.492 \text{ rad}$

We use the cosine law to find the displacement:

$\Delta x^2 = r^2 + r^2 - 2r^2 \cos\theta$

$\Delta x = \sqrt{2(20.0 \text{ m})^2(1 - \cos(0.492))} = 9.74 \text{ m}$

69. The throw angle for the close monkey is

$$\theta_c = \tan^{-1}\left(\frac{7.0}{7.0}\right) = 45.0°$$

For the far monkey we have

$$\theta_f = \tan^{-1}\left(\frac{7.0}{13}\right) = 28.3°$$

Since the stone thrown by the close monkey takes 3.0 s to reach the ground, we can calculate v:

$$y_c = v \sin(45.0°)(3.0\text{ s}) - \frac{1}{2}(9.81\text{ m/s}^2)(3.0\text{ s})^2 = 0$$

$$v = 20.81\text{ m/s}$$

Let us now work in the frame of reference of the stone thrown by the close monkey. The velocity of the other stone has the following components:

$$v_x = 2 \times 20.81\text{ m/s} \times \cos(28.3°) + 20.81\text{ m/s} \times \cos(45.0°) = 51.4\text{ m/s}$$
$$v_y = 2 \times 20.81\text{ m/s} \times \sin(28.3°) - 20.81\text{ m/s} \times \sin(45.0°) = 5.02\text{ m/s}$$

Relative to the stone tossed by the close monkey, the other stone will move toward it at a constant velocity, because the relative acceleration is zero. The displacement and velocity are in the same direction, so we can find the distance above the close monkey's stone:

$$\frac{\Delta y}{20\text{ m}} = \frac{5.02\text{ m/s}}{51.4\text{ m/s}}$$

$$\Delta y = 2.0\text{ m}$$

71. (a) The particle is at $x = 20$, $y = 0$ at $t = 0$.

By differentiating the formulas in problem 70, we have

$$v_x(t) = -8.00\text{ m/s} \times \sin(0.400 \times t)$$
$$v_y(t) = 8.00\text{ m/s} \times \cos(0.400 \times t)$$
$$\vec{v}(12.0\text{ s}) = (7.97\text{ m/s})\hat{i} + (0.70\text{ m/s})\hat{j}$$

(b) The average velocity is given by

$$\vec{v}_{avg} = \frac{\vec{r}(20.0\text{ s}) - \vec{r}(12.0\text{ s})}{20.0\text{ s} - 12.0\text{ s}}$$

$$= \frac{(20.0\text{ m})(\cos(0.40 \times 20.0)\hat{i} + \sin(0.40 \times 20.0)\hat{j}) - (20.0\text{ m})(\cos(0.40 \times 12.0)\hat{i} + \sin(0.40 \times 12.0)\hat{j})}{8.0\text{ s}}$$

$$= (-0.582\text{ m/s})\hat{i} - (4.96\text{ m/s})\hat{j}$$

For the average acceleration,

$$\vec{v}(20.0 \text{ s}) = (-7.91 \text{ m/s})\hat{i} - (1.16 \text{ m/s})\hat{j}$$

$$\vec{a}_{avg} = \frac{\vec{v}(20.0 \text{ s}) - \vec{v}(12.0 \text{ s})}{20.0 \text{ s} - 12.0 \text{ s}} = \frac{(-7.91 \text{ m/s})\hat{i} - (1.16 \text{ m/s})\hat{j} - ((7.97 \text{ m/s})\hat{i} + (0.700 \text{ m/s})\hat{j})}{8.0 \text{ s}}$$

$$= (-1.98 \text{ m/s}^2)\hat{i} - (0.23 \text{ m/s}^2)\hat{j}$$

73. Find the velocity of the small boat with respect to the rescue boat. The rescue boat has a speed of 30. km/h and a bearing of θ, measured in degrees north of east:

$$v_{sr_x} = 13 \text{ km/h} \times \cos(37°) - 30 \text{ km/h} \times \cos(\theta)$$
$$v_{sr_y} = 13 \text{ km/h} \times \sin(37°) - 30 \text{ km/h} \times \sin(\theta)$$

The rescue boat should see the small boat heading directly toward it. Therefore, the direction should be 42° south of east:

$$\frac{v_{sr_y}}{v_{sr_x}} = \tan(-42°)$$

$$\frac{7.82 \text{ km/h} - (30 \text{ km/h})\sin\theta}{10.38 \text{ km/h} - (30 \text{ km/h})\cos\theta} = -0.900$$

This was solved numerically, giving $\theta = 113°$. The rescue ship must head 23° west of north to intercept the small boat.

75. (a) $v_{avg} = \frac{\Delta \text{Altitude}}{\Delta t} = \frac{200 \text{ km}}{550 \text{ s}} = 0.36 \text{ km/s}$

(b) We need to estimate the slopes of the line at these times. By inspection, they are 0.50 km/s, 0.25 km/s, and 2.5 km/s.

(c) Altitude versus time increases linearly during this interval, suggesting that the acceleration is zero.

(d) Since the graph shows altitude versus time, the slope does not reflect any horizontal component of the velocity.

77. The horizontal distance the ball must cover to clear the tree is

$$v\cos(47.0°)t = 40.0 \text{ m} \tag{1}$$

The vertical distance the ball must cover the clear the tree is

$$v\sin(47.0°)t - \frac{1}{2}(9.81 \text{ m/s}^2) \times t^2 = 12.0 \text{ m} \tag{2}$$

Solving (1) and (2) simultaneously gives $v = 23.4$ m/s. The ball will clear the tree with any velocity greater than 23.4 m/s. The result is that the ball will clear the tree by greater and greater distances.

Chapter —FLUIDS

1. c. As the gas cools, its volume decreases. Since density is inversely proportional to volume, the density increases.

3. d. The net force at the bottom of the cylinder is proportional to the area of the bottom.

5. A, B, C. The density corresponds to the slope of the graphed line.

7. a. When the iron cube is in the boat, the volume of water displaced has the same weight as the weight of the cube. However, when the cube is dropped in the water, the volume of water displaced is equal to the volume of the cube. The first volume of water is greater than the second volume of water. Therefore, the water level will fall.

9. c. The magnitude of the buoyant force on a floating object is equal to the weight of the object. In other words, the floating object displaces an amount of water equal to its weight. When the ice cube melts, the amount of extra water in the glass is exactly the same as the amount of water displaced, so the water level remains the same.

11. b. The scale measures the weight of the container and its contents. If instead of placing the ball in the container, an amount of water with weight equal to the ball's weight were placed in the container, the scale would read the same in each case.

13. Yes; the pressure at the bottom depends on the height of the water, not on the surface area at the bottom.

15. a. The tension on the string is given by

$$T = Mg - F_B = \rho_b V_b g - \rho_f V_b g$$

Given that $\rho_b = 4\rho_f$,

$$T = \rho_b V_b g - \frac{\rho_b}{4} V_b g = \frac{3}{4} \rho_b V_b g = \frac{3}{4} Mg$$

We have assumed that given densities are exact when calculating significant digits in these solutions.

17. The number of water molecules in 1.0 cm³ is

$$N = \frac{6.02 \times 10^{23}}{18 \text{ g}} (1.0 \text{ g/cm}^3) = 3.344 \times 10^{22} / \text{cm}^3$$

Assume each molecule to be at the centre of a cube of length x. The volume of each cube is

$$x^3 = \frac{1}{3.344 \times 10^{22}} \text{ cm}^3 = 3.00 \times 10^{-23} \text{ cm}^3$$

$$x = 3.1 \times 10^{-8} \text{ cm}$$

Since the molecules are at the centre of these cubes, the distance between the molecules is also 3.1×10^{-8} cm = 0.31 nm.

19. $P = \dfrac{F}{A} = \dfrac{(70 \text{ kg})(9.81 \text{ m/s}^2)}{0.010 \text{ m}^2} = 68670 \text{ N/m}^2 = 69 \text{ kN/m}^2$

21. (a) $\rho = \dfrac{M}{\frac{4}{3}\pi R^3} = \dfrac{3.00 \times 10^{30} \text{ kg}}{\frac{4}{3}\pi(1.20 \times 10^3 \text{ m})^3} = 4.14 \times 10^{20} \text{ kg/m}^3$

    (b) $W = mg = \rho V g$
    $= (4.14 \times 10^{20} \text{ kg/m}^3)(1.0 \times 10^{-6} \text{ m}^3)(9.81 \text{ m/s}^2) = 4.1 \times 10^{15}$ N

    It would be impossible for one person to lift such a heavy weight.

23. The volume of the solid portion of the shell is

    $V = \dfrac{4}{3}\pi r_2^3 - \dfrac{4}{3}\pi r_1^3$
    $= \dfrac{4}{3}\pi(r_2^3 - r_1^3) = \dfrac{4}{3}\pi[(0.15 \text{ m})^3 - (0.14 \text{ m})^3] = 0.002\,64 \text{ m}^3$

    Therefore, the mass of the solid portion is

    $M = \rho V = (6000 \text{ kg/m}^3)(0.002\,64 \text{ m}^3) = 15.84$ kg

    The average density of the shell, including the hollow portion (assuming the hollow is evacuated), is given by

    $\rho_{avg} = \dfrac{M}{\frac{4}{3}\pi r_2^3} = \dfrac{15.84 \text{ kg}}{\frac{4}{3}\pi(0.15 \text{ m})^3} = 1120 \text{ kg/m}^3$

    Since the average density is greater than the density of water, the shell will not float in water.

25. The force on the window from inside the submarine is

    $F_{in} = P_0 A = (1.013 \times 10^5 \text{ Pa})[\pi \times (0.2 \text{ m})^2] = 12\,700$ N

    The pressure on the window from outside the submarine is

    $P = \rho g h + P_0 = (1030 \text{ kg/m}^3)(9.81 \text{ m/s}^2)(45.0 \text{ m}) + 101\,300 \text{ Pa} = 5.56 \times 10^5$ Pa

    Therefore, the force on the window from outside the submarine is

    $F_{out} = PA = (5.56 \times 10^5 \text{ Pa})[\pi \times (0.2 \text{ m})^2] = 69\,900$ N

    Hence, the net force on the window is

    $F_{net} = F_{out} - F_{in} = 69\,900 \text{ N} - 12\,700 \text{ N} = 57\,200 \text{ N} = 57$ kN

27. $P = P_0 + \rho g h$

   $= (1.013 \times 10^5 \text{ Pa}) + (1000 \text{ kg/m}^3)(9.81 \text{ m/s}^2)(0.9 \text{ m} - 0.3 \text{ m}) = 107 \text{ kPa}$

29.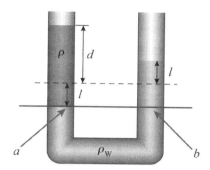

Consider a horizontal line that passes through the intersection of the two fluids. Consider two points, a and b, along this line, as shown in the figure. Let P_a and P_b be the fluid pressure at these points. Since the line is horizontal and it passes through the body of the same static fluid,

$P_a = P_b$

Hence,

$P_0 + \rho g(d+l) = P_0 + \rho_w g 2l$

This gives

$\rho(d+l) = \rho_w 2l$

$\dfrac{\rho}{\rho_w} = \dfrac{2l}{d+l}$

31. a. The total area of the output arms in the second press is about twice that of the first, but so is the force exerted on each output arm. Thus, the pressure is the same at each output arm, and so the pressure is also the same at each input arm, by Pascal's principle.

33. $\dfrac{V_{in}}{V_o} = \dfrac{\rho_{obj}}{\rho_{fl}}$

   $\dfrac{V_{in}}{V_o} = \dfrac{0.85}{1.2}$

   $V_{in} = 0.71 V_o$

35. The total force on the object is

    $F = 500.0 \text{ N} + 30.0 \text{ N} = 530.0 \text{ N}$

    This force is equal to the force of buoyancy on the object:

    $\rho_w g V = 530.0 \text{ N}$

    $V = \dfrac{530.0 \text{ N}}{\rho_w g} = \dfrac{530.0 \text{ N}}{(1000 \text{ kg/m}^3)(9.81 \text{ m/s}^2)} = 0.0540 \text{ m}^3$

    Hence, the density of the object is

    $\rho = \dfrac{M}{V} = \dfrac{500.0 \text{ N}}{(9.81 \text{ m/s}^2)(0.054 \text{ m}^3)} = 944 \text{ kg/m}^3$

37. (a) For the first sphere, we have

    $\dfrac{V_{in}}{V_o} = \dfrac{\rho_1}{\rho_{fl}} = \dfrac{1}{2}$

    $\dfrac{m}{V \rho_{fl}} = \dfrac{1}{2}$ \quad (1)

    Similarly, for the second sphere, we have

    $\dfrac{2m}{2V \rho_{fl}} = h$ \quad (2)

    Dividing (2) by 2 and substituting into (1) gives $h = 1/2$. Hence, the second sphere will be half submerged in the water as well.

    (b) Since the first sphere is 50% under water, we have

    $\dfrac{\rho_0}{\rho_w} = \dfrac{1}{2}$

    $\dfrac{m}{V} = \dfrac{1}{2} \rho_w$

    $V = \dfrac{2m}{\rho_w}$

The buoyant force is therefore given by

$$F_B = \rho_w g V'$$

where V' is the volume of water displaced. But

$$V' = \frac{V}{2} = \frac{m}{\rho_w}$$

$$F_B = mg$$

(c) Let m and V represent the mass and volume of the small sphere, and let m_2 and V_2 represent the mass and volume of the large sphere. The second sphere is also 50% under water, so we have

$$\frac{\rho_0}{\rho_w} = \frac{1}{2}$$

$$\frac{m_2}{V_2} = \frac{1}{2}\rho_w$$

$$V_2 = \frac{m_2}{2\rho_w}$$

The buoyant force is therefore given by

$$F_B = \rho_w g V'$$

where V' is the volume of water displaced. But

$$V' = \frac{V_2}{2} = \frac{m_2}{\rho_w}$$

$$F_B = m_2 g = 2mg$$

(d) The buoyant force on the second sphere is twice as large as the buoyant force on the first sphere, as shown in part (c).

(e) The volume of water displaced by the second sphere is twice as large as the volume of water displaced by the first sphere.

39. The volume of the hydrometer submerged in water is

$$V_{sub,w} = 13.2 \times 10^{-6} \text{ m}^3 - (0.40 \times 10^{-4} \text{ m}^2)(0.080 \text{ m}) = 1.0 \times 10^{-5} \text{ m}^3$$

Since

$$\frac{V_{sub,w}}{V_0} = \frac{\rho_{hyd}}{\rho_{water}}$$

we have

$$\rho_{hyd} = \frac{V_{sub,w}}{V_0}\rho_{water} = \frac{1.0\times 10^{-5} \text{ m}^3}{13.2\times 10^{-6} \text{ m}^3}(1000 \text{ kg/m}^3) = 758 \text{ kg/m}^3$$

Now, the volume of the hydrometer submerged in alcohol is

$$V_{sub,alc} = 13.2\times 10^{-6} \text{ m}^3 - (0.40\times 10^{-4} \text{ m}^2)(0.010 \text{ m}) = 1.28\times 10^{-5} \text{ m}^3$$

Since

$$\frac{V_{sub,alc}}{V_0} = \frac{\rho_{hyd}}{\rho_{alc}}$$

we have

$$\rho_{alc} = \frac{V_0}{V_{sub,alc}}\rho_{hyd} = \frac{13.2\times 10^{-6} \text{ m}^3}{1.28\times 10^{-5} \text{ m}^3}(758 \text{ kg/m}^3) = 780 \text{ kg/m}^3$$

41. The total volume of the cube is $V_t = (20 \text{ cm})^3 = 8000 \text{ cm}^3$. The volume of the inside hollow portion of the cube is $V_h = (20 - 2x)^3$, where x is the thickness of the aluminum. Hence, the volume of the solid part of the cube, V_s, is given by $V_s = V_t - V_h = (8000 - (20 - 2x)^3) \text{ cm}^3$. The mass of the solid part (in grams), is given by

$$m_s = \rho_s V_s = 2.7 \text{ g/cm}^3 \times [8000 \text{ cm}^3 - (20-2x)^3]$$

Assume that the mass of the air inside the cube is negligible. This implies that the mass of aluminum is the total mass of the cube:

$$m_t = m_s = 2.7 \text{ g/cm}^3 \times [8000 \text{ cm}^3 - (20-2x)^3]$$

Since we need the average density to be 1 g/cm³, the total mass of the cube must be equal to (1 g/cm³)(8000 cm³) = 8000 g:

$$2.7 \text{ g/cm}^3 \times [8000 \text{ cm}^3 - (20-2x)^3] = 8000 \text{ g}$$
$$x = 1.43 \text{ cm}$$

43. The mass of the spherical shell is given by

$$m = (2700 \text{ kg/m}^3)\left(\frac{4}{3}\pi(0.5000 \text{ m})^3 - \frac{4}{3}\pi r_i^3\right)$$

Since the sphere is completely immersed in the water, we have

$$\frac{\rho_{sphere}}{\rho_{water}} = 1$$

$$\rho_{sphere} = \rho_{water}$$

$$\frac{(2700 \text{ kg/m}^3)\left(\frac{4}{3}\pi(0.5000 \text{ m})^3 - \frac{4}{3}\pi r_i^3\right)}{\frac{4}{3}\pi(0.5000 \text{ m})^3} = 1000 \text{ kg/m}^3$$

$$\frac{(2700 \text{ kg/m}^3)[(0.5000 \text{ m})^3 - r_i^3]}{(0.5000 \text{ m})^3} = 1000 \text{ kg/m}^3$$

$$\frac{(0.5000 \text{ m})^3 - r_i^3}{(0.5000 \text{ m})^3} = \frac{1000}{2700}$$

$$(0.5000 \text{ m})^3 - r_i^3 = \frac{(0.5000 \text{ m})^3}{2.7}$$

$$r_i^3 = (0.5000 \text{ m})^3 - \frac{(0.5000 \text{ m})^3}{2.7}$$

$$r_i = 0.429 \text{ m} = 43 \text{ cm}$$

Thus, the inner diameter of the shell is about 86 cm. If there were air in the hollow, then the mass of the aluminum would be a little less, and so its thickness would also be a little less.

45. (a) $\dfrac{V_{sub}}{V_0} = \dfrac{\rho_{wood}}{\rho_{water}} = \dfrac{600 \text{ kg/m}^3}{1000 \text{ kg/m}^3} = 0.60$

(b) Force balance gives

$$W_{iron} + W_{block} = F_B$$
$$W_{iron} = \rho_w g V_{sub} - W_{block}$$

The volume of the block is

$$V_b = \frac{m_b}{\rho_b} = \frac{W_b/g}{\rho_b} = \frac{4.0 \text{ N}}{(9.81 \text{ m/s}^2)(600 \text{ kg/m}^3)} = 6.8 \times 10^{-4} \text{ m}^3$$

It is given that $V_{sub}/V_b = 0.85$. Thus,

$$V_{sub} = (0.85)(6.8 \times 10^{-4} \text{ m}^3) = 5.78 \times 10^{-4} \text{ m}^3$$

Hence, the weight of the iron is

$$W_{iron} = (1000 \text{ kg/m}^3)(9.81 \text{ m/s}^2)(5.78 \times 10^{-4} \text{ m}^3) - 4.0 \text{ N} = 1.67 \text{ N} = 1.7 \text{ N}$$

(c) The volume of the iron is

$$V_{iron} = \frac{m_{iron}}{\rho_{iron}} = \frac{W_{iron}/g}{\rho_{iron}} = \frac{1.67 \text{ N}}{(9.81 \text{ m/s}^2)(6000 \text{ kg/m}^3)} = 2.83 \times 10^{-5} \text{ m}^3$$

The effective density is then given by

$$\rho = \frac{4.0 \text{ N} + 1.67 \text{ N}}{(9.81 \text{ m/s}^2)(6.8 \times 10^{-4} \text{ m}^3 + 2.83 \times 10^{-5} \text{ m}^3)} = 816 \text{ kg/m}^3$$

$$\frac{V_{sub}}{V_0} = \frac{\rho}{\rho_{water}} = \frac{816 \text{ kg/m}^3}{1000 \text{ kg/m}^3} = 0.82$$

47. (a) For an object floating in a fluid,

    $$\frac{V_{immersed}}{V_{fluid}} = \frac{\rho_{object}}{\rho_{fluid}}$$

    where ρ_{object} is the average density of the object and $V_{immersed}$ is the volume of the object immersed in the fluid. For the given situation,

    $$\frac{V_{immersed}}{V_{fluid}} = \frac{800 \text{ kg/m}^3}{1000 \text{ kg/m}^3} = 0.8$$

    (b) Let ΔF_B be the additional buoyant force exerted on the block when the remaining 20% of its volume submerges in water. Then,

    $$\Delta F_B = (0.2 \times V_{block})\rho_{water}g = 0.2 \times (0.3 \text{ m} \times 0.1 \text{ m} \times 0.05 \text{ m}) \times 1000 \text{ kg/m}^3 \times 9.8 \text{ m/s}^2 = 0.3g$$

    The weight of the aluminum piece placed on top of the block must then be equal to this additional buoyant force. Therefore,

    $$\Delta F_B = m_{al}g$$

    $$m_{al} = \frac{\Delta F_B}{g} = \frac{0.3g}{g} = 0.3 \text{ kg}$$

    (c) If the aluminum piece is to be glued to the bottom of the block, then there will be a buoyant force exerted on it as well. So the mass of the piece would have to be slightly greater (compared to the case when the aluminum piece is on top of the block and hence does not submerge in the water) to overcome this additional buoyant force.

49. When the cylinder is immersed in the water, it experiences a buoyant force of magnitude equal to that of the weight of the water it displaces. The increase in the reading on the weigh scale is equal to the weight of the water displaced by the cylinder (assuming that the water does not spill). Since

    $$F_B = (\text{volume of immersed cylinder}) \times \rho_{water}g$$
    $$= \pi(1.0 \times 10^{-2} \text{ m})^2(4.0 \times 10^{-2} \text{ m}) \times 1000 \text{ kg/m}^3 \times 9.8 \text{ m/s}^2 = 0.12 \text{ N}$$

    the reading on the scale is

    $$4.4 \text{ N} + 0.12 \text{ N} = 4.5 \text{ N}$$

As the buoyant force and the weight of the cylinder act vertically and in opposite directions,

$T = W_{cylinder} - F_B$

where T is the magnitude of the tension in the string. Therefore,

$T = \pi(1.0 \times 10^{-2} \text{ m})^2 (8.0 \times 10^{-2} \text{ m}) \times 2700 \text{ kg/m}^3 \times 9.8 \text{ m/s}^2 - 0.12 \text{ N}$
$= 0.664 \text{ N} - 0.12 \text{ N} = 0.54 \text{ N}$

Thus, the reading on the scale, $W_{cylinder}$, is 0.66 N, and the tension in the string is 0.54 N.

51. $A_1 v_1 = A_2 v_2$

$v_2 = \dfrac{A_1 v_1}{A_2} = \dfrac{\pi r_1^2 v_1}{\pi r_2^2} = \dfrac{r_1^2 v_1}{r_2^2}$

$v_2 = \left(\dfrac{0.020 \text{ m}}{0.010 \text{ m}}\right)^2 (0.1 \text{ m/s}) = 0.4 \text{ m/s}$

53. The volume of the balloon is given by

$V = \dfrac{4}{3}\pi r^3$

Differentiating with respect to time gives

$\dfrac{dV}{dt} = 4\pi r^2 \dfrac{dr}{dt}$

$\dfrac{dr}{dt} = \dfrac{1}{4\pi r^2}\dfrac{dV}{dt} = \dfrac{1}{4\pi (0.2 \text{ m})^2}(0.1 \times 10^{-3} \text{ m}^3/\text{s}) = 2 \times 10^{-4} \text{ m/s} = 0.2 \text{ mm/s}$

55. According to Bernoulli's principle,

$P_1 + \dfrac{1}{2}\rho v_1^2 + \rho g y_1 = P_2 + \dfrac{1}{2}\rho v_2^2 + \rho g y_2$

Because the speed in the main is negligible, we can set $v_1 = 0$. Thus,

$P_1 + \rho g y_1 = P_2 + \dfrac{1}{2}\rho v_2^2 + \rho g y_2$

$v_2^2 = \dfrac{2}{\rho}[(P_1 - P_2) + (\rho g y_1 - \rho g y_2)] = \dfrac{2}{\rho}[(P_1 - P_2) + \rho g(y_1 - y_2)] = \dfrac{2}{\rho}[(P_1 - P_2) + 15.0\rho g]$

Here P_1 and P_2 are the total pressures:
$P_1 = 170\,000 \text{ Pa} + P_0$
$P_2 = P_0$

where P_0 is the atmospheric pressure. The maximum speed occurs if the gauge pressure on the fifth floor is negligible:

$$v_2^2 = \frac{2}{\rho}[(170\,000\text{ Pa}) - (15\text{ m})\rho g] = \frac{340\,000\text{ Pa}}{1000\text{ kg/m}^3} - (15\text{ m})g$$

$$v_2 = \sqrt{340\text{ m}^2/\text{s}^2 - (15\text{ m})(9.81\text{ m/s}^2)} = 14\text{ m/s}$$

57. Assuming the height does not change, Bernoulli's principle gives

$$P_1 + \frac{1}{2}\rho v_1^2 = P_2 + \frac{1}{2}\rho v_2^2$$

$$v_2 = \sqrt{\frac{2}{\rho}(P_1 - P_2) + v_1^2}$$

Given that $P_1 - P_2 = 7.0 \times 10^3$ Pa,

$$v_2 = \sqrt{\frac{2}{1000\text{ kg/m}^3}(7.0 \times 10^3\text{ Pa}) + 0.0025\text{ m}^2/\text{s}^2} = 3.7\text{ m/s}$$

59. (a) $A_1 v_1 = A_2 v_2$

$$v_2 = \frac{A_1 v_1}{A_2} = \frac{(4.00\text{ cm}^2)(0.5\text{ m/s})}{0.50\text{ cm}^2} = 4.0\text{ m/s}$$

(b) According to Bernoulli's principle,

$$P_1 + \rho g y_1 + \frac{1}{2}\rho v_1^2 = P_2 + \rho g y_2 + \frac{1}{2}\rho v_2^2$$

$$P_1 = P_2 + \rho g(y_2 - y_1) + \frac{1}{2}\rho(v_2^2 - v_1^2)$$

$$P_1 = 1.01 \times 10^5\text{ Pa} + (1000\text{ kg/m}^3)(9.81\text{ m/s}^2)(1.0\text{ m})$$

$$+ \frac{1}{2}(1000\text{ kg/m}^3)[(4.0\text{ m/s})^2 - (0.5\text{ m/s})^2]$$

$$= 120\text{ kPa}$$

61. (a) According to Bernoulli's equation,

$$P_1 + \rho g y_1 + \frac{1}{2}\rho v_1^2 = P_2 + \rho g y_2 + \frac{1}{2}\rho v_2^2$$

Since $P_1 = P_2$ and $v_1 \approx 0$ due to the large area of the lake,

$$v_2 = \sqrt{2g(y_1 - y_2)} = \sqrt{2(9.81\text{ m/s}^2)(20.0\text{ m})} = 19.8\text{ m/s}$$

Since $A_2v_2 = A_ev_e$,

$$v_e = \frac{A_2v_2}{A_e} = \left(\frac{r_2}{r_e}\right)^2 v_2 = \left(\frac{0.40 \text{ m}}{0.10 \text{ m}}\right)^2 (19.8 \text{ m/s}) = 317 \text{ m/s}$$

(b) The total pressure inside the pipe is given by

$$P = P_{atm} + \rho g \Delta y + \frac{1}{2}\rho v_2^2$$

$$= 1.01 \times 10^5 \text{ Pa} + (1000 \text{ kg/m}^3)(9.81 \text{ m/s}^2)(20.0 \text{ m}) + \frac{1}{2}(1000 \text{ kg/m}^3)(19.8 \text{ m/s})^2$$

$$= 4.93 \times 10^5 \text{ Pa} = 493 \text{ kPa}$$

63.
$$A_1v_1 = A_2v_2$$
$$\pi(0.015 \text{ m})^2 v_1 = \pi(0.0050 \text{ m})^2 v_2$$
$$v_2 = 9v_1$$

Now, according to Bernoulli's theorem,

$$P_1 + \rho_{air}gy_1 + \frac{1}{2}\rho_{air}v_1^2 = P_2 + \rho_{air}gy_2 + \frac{1}{2}\rho_{air}v_2^2$$

Since $y_1 = y_2$,

$$v_2^2 - v_1^2 = \frac{2}{\rho_{air}}(P_1 - P_2)$$

Given $P_1 - P_2 = \Delta P = \rho_m g \Delta y = (13\,600 \text{ kg/m}^3)(9.81 \text{ m/s}^2)(0.0010 \text{ m}) = 133$ Pa, we have

$$v_2^2 - v_1^2 = \frac{2}{1.29 \text{ kg/m}^3}(133 \text{ Pa}) = 206 \text{ m}^2/\text{s}^2$$

where we have used $\rho_{air} = 1.29$ kg/m^3. Since $v_2 = 9v_1$,

$$(9v_1)^2 - v_1^2 = 206 \text{ m}^2/\text{s}^2$$
$$80v_1^2 = 206 \text{ m}^2/\text{s}^2$$
$$v_1 = 1.6 \text{ m/s}$$
$$v_2 = 9 \times 1.6 = 14 \text{ m/s}$$

65. According to Bernoulli's theorem

$$P_1 + \rho gy_1 + \frac{1}{2}\rho v_1^2 = P_2 + \rho gy_2 + \frac{1}{2}\rho v_2^2$$

Since the tank is very large, $v_1 = 0$. Also, $y_2 = 0$. Given that $P_1 = 3P_0$ and $P_2 = P_0$, we have

$$3P_0 + \rho g y_1 = P_0 + \frac{1}{2}\rho v_2^2$$

$$v_2 = \sqrt{\frac{2}{\rho}(2P_0 + \rho g y_1)}$$

$$= \sqrt{\frac{2}{1000 \text{ kg/m}^3}[2 \times 1.01 \times 10^5 \text{ Pa} + (1000 \text{ kg/m}^3)(9.81 \text{ m/s}^2)(3 \text{ m})]} = 21.5 \text{ m/s}$$

Hence, the flow rate is given by

$$Q = Av_2 = (2.0 \times 10^{-4} \text{ m}^2)(21.5 \text{ m/s}) = 0.0043 \text{ m}^3/\text{s} = 4.3 \text{ L/s}$$

67.

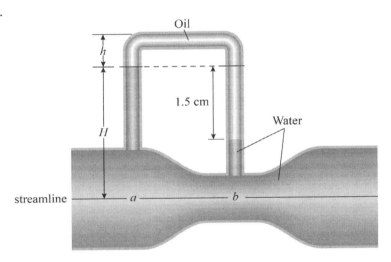

Consider points a and b on a horizontal streamline that passes through the centre of the flowmeter, as shown in the figure. Since we know the cross-sectional areas of the tube at positions a and b, we can determine the relation between the flow speeds at these points by applying the continuity equation:

$$A_a v_a = A_b v_b$$

which gives

$$\pi(1.0 \text{ cm})^2 v_a = \pi(0.40 \text{ cm})^2 v_b$$

$$v_a = 0.16 v_b \qquad (1)$$

But we do not know v_a or v_b. Let us now apply Bernoulli's equation at a and b. As the flow is horizontal, the potential energy per unit volume term drops out. Therefore,

$$P_a + \frac{1}{2}\rho v_a^2 = P_b + \frac{1}{2}\rho v_b^2$$

Therefore,

$$P_a - P_b = \frac{1}{2}\rho(v_b^2 - v_a^2) \qquad (2)$$

We can determine the pressure difference $P_a - P_b$ by determining the hydrostatic pressures at these points due to the columns of water and oil above these positions. Let H be the height of the water column in the vertical tube that is above position a and h be the height of the column of oil above the water–oil interface in this tube. Then,

$$P_a = \rho_{water}gH + \rho_{oil}gh$$

Similarly,

$$P_b = \rho_{water}g(H - 1.5\times 10^{-2} \text{ m}) + \rho_{oil}g(h + 1.5\times 10^{-2} \text{ m})$$

Therefore,

$$P_a - P_b = (\rho_{water} - \rho_{oil})\times g \times 1.5\times 10^{-2} \text{ m}$$
$$= 200 \text{ kg/m}^3 \times 9.8 \text{ m/s}^2 \times 1.5\times 10^{-2} \text{ m}$$
$$= 29.4 \text{ Pa}$$

Inserting this value in (2) and using (1), we get

$$29.4 \text{ Pa} = \frac{1}{2}(1000 \text{ kg/m}^3)\times[v_b^2 - (0.16)^2 v_a^2]$$

Solving for v_b, we get

$$v_b = 24.6 \text{ cm/s}$$

Therefore, the flow rate through the tube is

$$Q = A_b v_b = \pi(1.0 \text{ cm})^2 \times 24.6 \text{ cm/s} = 77 \text{ cm}^3/\text{s}$$

69. (a) Pressure in the radial direction is given by

$$p = \frac{1}{2}\rho v^2$$

Using $v = r\omega$, we get

$$p = \frac{1}{2}\rho r^2 \omega^2$$

Differentiating with respect to r gives

$$\frac{dp}{dr} = \rho r \omega^2$$

(b) $\dfrac{dp}{dr} = \rho r \omega^2$

$dp = \rho r \omega^2 dr$

Integrating both sides, we get

$\displaystyle\int_{P_0}^{P_R} dp = \int_0^R \rho r \omega^2 dr$

$P_R - P_0 = \rho \omega^2 \dfrac{R^2}{2}$

$P_R = P_0 + \dfrac{1}{2}\rho \omega^2 R^2$

(c) Consider a particle of mass m in the rotating frame of reference of the fluid. In this frame, the fluid is in static equilibrium. For simplicity, we will look at the particle in the xy-frame only and then extend the result to the third dimension. The centripetal force acting on the particle is in the x-direction and is given by

$F_c = m\omega^2 x$

The gravitational force is in the y-direction and is given by

$F_g = mg$

The angle the resultant force makes with the horizontal is given by

$\tan\theta = \dfrac{F_c}{F_g} = \dfrac{\omega^2 x}{g}$

But $\tan\theta = dy/dx$, so

$dy = \dfrac{\omega^2 x}{g} dx$

Integrating both sides, we get

$y = \dfrac{\omega^2 x^2}{2g} + C$

This is the equation of a parabola. In three dimensions, this will form a paraboloid.

71. Given that $A = 1.0 \times 10^{-4}$ m^2, we have

$R^4 = \left(\dfrac{A}{\pi}\right)^2 = \left(\dfrac{1.0 \times 10^{-4} \text{ m}^2}{\pi}\right)^2 = 1.013 \times 10^{-9}$ m^4

The pressure difference is given by

$$\Delta P = \frac{8\mu LQ}{\pi R^4} = \frac{8(1.00\times 10^{-3}\text{ Pa}\cdot\text{s})(5.0\text{ m})(5.0\times 10^{-4}\text{ m}^3/\text{s})}{\pi(1.013\times 10^{-9}\text{ m}^4)} = 6.28\times 10^3\text{ Pa} = 6.3\text{ kPa}$$

73. From Poiseuille's equation, the pressure difference required to maintain a flow rate of Q through a pipe of radius R and length L is given by

$$P_1 - P_2 = \frac{8\mu LQ}{\pi R^4}$$

Therefore, for a given pressure difference, the viscosity of the fluid is given by

$$\mu = \frac{(P_1 - P_2)\pi R^4}{8LQ}$$

Inserting the given values,

$$\mu = \frac{\pi(5\times 10^4\text{ Pa})\times(5.0\times 10^{-2}\text{ m})^4}{8(200\text{ m})\times(0.2\text{ m}^3/\text{s})} = 3\times 10^{-3}\text{ Pa}\cdot\text{s}$$

75. The diameter of the needle is small, so the viscosity of the blood cannot be ignored and we must use Poiseuille's equation to determine the pressure difference needed to maintain the required flow rate. Let P_{arm} be the pressure at the end of the needle that is inserted in the patient's arm and P_{pull} be the pressure that needs to be maintained to pull the plunger outward. We will assume that the pressure at the end of the needle that is closer to the plunger is P_{pull}.

The following quantities are given:

Radius of the needle, $R = 0.50$ mm/2 $= 0.25 \times 10^{-3}$ m

Length of the needle, $L = 6.0$ cm $= 6.0 \times 10^{-2}$ m

Volume flow rate of the blood through the needle,

$Q = 0.10$ cm^3/s $= 0.10 \times (10^{-2}$ m$)^3$/s $= 1.0 \times 10^{-7}$ m^3/s

Blood viscosity, $\mu = 3.5 \times 10^{-4}$ Pa·s

Pressure at the end of the needle that is inserted in the arm, $P_{arm} = 85$ mm Hg

Since 760 mm Hg $= 1.0$ atmospheric pressure $= 1.01 \times 10^5$ Pa, 85 mm Hg $= 11\,300$ Pa.

P_{pull} can be calculated from Poiseuille's equation, which for the current situation can be written as

$$P_{pull} = P_{arm} + 8\mu\frac{LQ}{\pi R^4}$$

$$= 11\,300\text{ Pa} + \frac{8\times(3.5\times 10^{-4}\text{ Pa}\cdot\text{s})\times(6.0\times 10^{-2}\text{ m})\times(1.0\times 10^{-7}\text{ m}^3/\text{s})}{\pi(0.25\times 10^{-3}\text{ m})^4}$$

$$= 11\,300\text{ Pa} + 1370\text{ Pa} = 12\,670\text{ Pa} = 13\text{ kPa to two significant digits}$$

77. (a) The apparent weight in water is given by

$$W_{app} = mg - \rho_w gV$$

$$V = \frac{mg - W_{app}}{\rho_w g} = \frac{790.0 \text{ N} - 50.0 \text{ N}}{(1000 \text{ kg/m}^3)(9.81 \text{ m/s}^2)} = 0.0754 \text{ m}^3$$

(b) $\rho = \frac{m}{V} = \frac{mg}{gV} = \frac{790.0 \text{ N}}{(9.81 \text{ m/s}^2)(0.0754 \text{ m}^3)} = 1068.0 \text{ kg/m}^3$

(c) Assuming that the total mass is the sum of the masses of fat and lean muscle, we have

$$m = m_F + m_L$$
$$\rho_P V = \rho_F V_F + \rho_L V_L \quad (1)$$

Let us also assume that the total volume is given by

$$V = V_F + V_L \quad (2)$$

The fraction of the total mass that is fat is given by

$$x_F = \frac{m_F}{m}$$
$$\rho_F V_F = x_F \rho_P V \quad (3)$$

Inserting (2) and (3) into (1) and simplifying, we get

$$\rho_P V = x_F \rho_P V + \rho_L V - x_F \frac{\rho_P \rho_L}{\rho_F} V$$

$$x_F = \frac{\rho_F}{\rho_P} \left(\frac{\rho_L - \rho_P}{\rho_L - \rho_F} \right)$$

(d) The lean and fat densities are the same throughout the body, and the entire body consists of just lean muscle and fat.

(e) $x_F = \frac{\rho_F}{\rho_P} \left(\frac{\rho_L - \rho_P}{\rho_L - \rho_F} \right) = \frac{900 \text{ kg/m}^3}{1068 \text{ kg/m}^3} \left(\frac{1100 \text{ kg/m}^3 - 1068 \text{ kg/m}^3}{1100 \text{ kg/m}^3 - 900 \text{ kg/m}^3} \right) = 0.135 = 13.5\%$

79. (a) Before the force is applied, the cube is in static equilibrium:

$$mg = \rho_w g V_{sub}$$

After the force, F, is applied, the volume submerges and the buoyant force increases, so

$$F + mg = \rho_w g V_{sub} + \rho_w g V'_{sub}$$

But $mg = \rho_w g V_{sub}$, so

$F = \rho_w g V'_{sub}$

If a side of the cube is L, its cross-sectional area is L^2, and $V'_{sub} = L^2 y$, where y is the additional submerged height after the force is applied:

$F = \rho_w g L^2 y$

$F \propto y$

(b) The magnitude of the force on the cube is given by

$F = \rho_w g L^2 y$

This can be written as

$F = ky$ with $k = \rho_w g L^2$

This is similar to the restoring force of a spring. Hence, water acts as a spring with spring constant $k = \rho_w g L^2$. Since this force is acting opposite to the direction of displacement of the cube, it should be written as $F = -ky$. The equation of motion is therefore given by

$m\ddot{y} + ky = 0$

$\ddot{y} + \dfrac{k}{m} y = 0$

where $\ddot{y}$ is the second derivative with respect to time. This is the equation of a simple harmonic oscillator.

(c) $\ddot{y} + \dfrac{k}{m} y = 0$

Thus,

$\omega = \sqrt{\dfrac{k}{m}} = \dfrac{2\pi}{T}$

$T = 2\pi \sqrt{\dfrac{m}{k}}$

The mass of the cube is given by $m = \rho_0 L^3$. We saw earlier that $k = \rho_w g L^2$. Thus,

$T = 2\pi \sqrt{\dfrac{\rho_0 L}{\rho_w g}}$

Chapter —OSCILLATIONS

1. The total distance travelled in one cycle by a simple harmonic oscillator is four times the amplitude, or $4A$. The oscillator reaches maximum positive amplitude A and maximum negative amplitude $-A$. Since the oscillator returns to the same position after one cycle, the net displacement after one cycle is zero.

3. For a mass–spring system, the period, T, is independent of the amplitude, A, because the period depends only on the angular frequency of the motion. This is evident by examining the equation of motion:

   $x(t) = A \cos(\omega t + \phi)$

   The period of a sinusoidal function is

   $T = \dfrac{2\pi}{\omega}$

   $= 2\pi \sqrt{\dfrac{m}{k}}$

   which is independent of the amplitude of oscillation.

5. If the mass of the spring in a mass–spring oscillator were not ignored, the period would be greater, because

   $T = 2\pi \sqrt{\dfrac{m}{k}}$

   If more mass is oscillating, the formula for the period indicates a greater period.

7. The period of the pendulum is greater if the mass of the string is not negligible, because then the pendulum is no longer a "simple" pendulum. The formula for the period of a physical pendulum includes the mass of the pendulum, and greater masses mean greater periods.

9. a. When the elevator accelerates upward, the effective value of g in the elevator increases. Since $T = 2\pi\sqrt{L/g}$ for a pendulum, the period will decrease.

11. The period of the swing decreases when the child stands up. If we approximate the standing child as a rod of length L, then the new period is $T = 2\pi\sqrt{\dfrac{ML^2/3}{MgL/2}} = 2\pi\sqrt{\dfrac{2}{3}}\sqrt{\dfrac{L}{g}}$, which is less than the period when the child sits and can be approximated as a point mass.

13. c. The period of a physical pendulum in air is greater than in vacuum because air resistance slows the pendulum.

15. d. Forces given by equations (a), (b), and (c) will not result in simple harmonic motion. For a particle to execute simple harmonic motion, the net force acting on it must be proportional to the particle's displacement from the equilibrium position (denoted by $x(t)$ in these equations) and opposite in sign. Only the force given by (d) satisfies this condition.

17. d. At the maximum amplitude of oscillation, the total energy, E_1, is equal to the potential energy, U, so $E_1 = \frac{1}{2}kA^2$. If the amplitude is doubled without "kicking" the oscillator, then the new energy is

$$E_2 = \frac{1}{2}k(2A)^2$$
$$= 4\left(\frac{1}{2}kA^2\right)$$
$$= 4E_1$$

If the amplitude is doubled, the total energy increases by a factor of 4.

19. Given quantities are $m = 2.0$ kg, $k = 5.0$ N/m, $x_i = 0.30$ m, $v_i = 0$ m/s.

    (a) The total energy of the mass–spring system is

    $$E = \frac{1}{2}mv^2 + \frac{1}{2}kx^2 = \frac{1}{2}\times 2.0\text{ kg}\times(0\text{ m/s})^2 + \frac{1}{2}\times 5.0\text{ N/m}\times(0.30\text{ m})^2 = 0.225\text{ J}$$

    Since $E = \frac{1}{2}kA^2$,

    $$A = \sqrt{\frac{2E}{k}} = \sqrt{\frac{2\times 0.225\text{ J}}{5.0\text{ N/m}}} = 0.30\text{ m}$$

    (b) At the maximum speed, the kinetic energy is equal to the total energy of the mass–spring system. Therefore,

    $$v_{max} = \sqrt{\frac{2E}{m}} = \sqrt{\frac{2\times 0.225\text{ J}}{2.0\text{ kg}}} = 0.47\text{ m/s}$$

    (c) Since $\omega = \sqrt{\frac{k}{m}}$, $\omega^2 = \frac{k}{m} = \frac{5.0\text{ N/m}}{2.0\text{ kg}} = 2.5\text{ (rad/s)}^2$

    $a_{max} = \omega^2 A = 2.5\text{ (rad/s)}^2 \times 0.30\text{ m} = 0.75\text{ m/s}^2$

    (d) Knowing the position and velocity at $t = 0$, we can determine the phase constant (see Example 13-3, equation (5)):

    $$\phi = \tan^{-1}\left(-\frac{v_i}{\omega x_i}\right) = \tan^{-1}(-0) = 0\text{ rad}$$

Therefore, since $\omega = \sqrt{2.5 \text{ (rad/s)}^2}$, the displacement as a function of time for the oscillator is given by $x(t) = (0.30 \text{ m})\cos(\sqrt{2.5}t)$.

21. (a) Comparing the given equation, $x(t) = (5 \text{ cm})\cos(2\pi t - \pi/4)$, with the standard equation for the position of a simple harmonic oscillator, $x(t) = A\cos(\omega t + \phi)$, we get

    $A = 5$ cm

    $\omega = 2\pi$ rad/s $\Rightarrow f = \dfrac{\omega}{2\pi} = \dfrac{2\pi \text{ rad/s}}{2\pi \text{ rad}} = 1$ Hz

    $\phi = -\pi/4$ rad

    (b) The equilibrium position occurs at $x = 0$. Therefore, the time t at which $x = 0$ is given by $\cos(2\pi t - \pi/4) = 0$. Therefore,

    $$2\pi t - \dfrac{\pi}{4} = \left(m + \dfrac{1}{2}\right)\pi$$

    $$t = \left(\dfrac{m}{2} + \dfrac{3}{8}\right) \text{ s}$$

    The time closest to $t = 0$ corresponds to $m = 0$, so

    $t = \dfrac{3}{8}$ s

    (c) $v(t) = \dfrac{d}{dt}(5 \text{ cm})\cos(2\pi t - \pi/4) = -(10\pi \text{ cm/s})\sin(2\pi t - \pi/4)$

    At $t = \dfrac{3}{8}$ s, we get

    $v(3/8 \text{ s}) = -(10\pi \text{ cm/s})\sin\left[2\pi\left(\dfrac{3}{8}\right) - \dfrac{\pi}{4}\right] = -10\pi$ cm/s

    (d) Since $v(3/8 \text{ s})$ is negative, the particle is moving in the direction of decreasing x.

    (e) $a_{max} = \omega^2 A = (2\pi \text{ rad/s})^2(5 \text{ cm}) = 20\pi^2$ cm/s^2

23. Because the mass is pulled to the left and then released from rest, we can take the phase angle to be π radians. Therefore, we can write its displacement from the equilibrium position as a function of time as

    $x(t) = A\cos(\omega t + \pi) = -A\cos(\omega t)$

    where A is the amplitude of oscillations. Differentiating with respect to time, we obtain

    $v(t) = \omega A \sin(\omega t)$

(a) Substituting the given information into the position and velocity equations, we obtain
$$0.10 \text{ m} = -A\cos(0.3\omega)$$
$$0.857 \text{ m/s} = \omega A \sin(0.3\omega)$$

Dividing the second equation by the first, we obtain
$$8.57 \text{ s}^{-1} \cos(0.3\omega) = -\omega \sin(0.3\omega)$$

By plotting the two sides of the equation on the same graph as a function of ω, we can look for the solutions where the curves intersect. The lowest value of ω at which the two curves intersect is
$$\omega = 7.67 \text{ rad/s}$$

so the frequency is
$$f = \frac{\omega}{2\pi} = 1.2 \text{ s}^{-1}$$

(b) $x(0.3 \text{ s}) = -A\cos(0.3\omega)$
$$0.10 \text{ m} = -A\cos[0.3(7.67)]$$
$$A = -\frac{0.10 \text{ m}}{\cos(2.30)}$$
$$A = -0.15 \text{ m}$$
$$A = -15 \text{ cm}$$

(c) The phase constant is π radians.

(d) $x(t) = -(0.15 \text{ m})\cos(7.7t + \pi) = -(0.15 \text{ m})\cos(7.7t)$

25. $x_1(t) = A\cos\omega t$

$x_2(t) = A\cos\left(\omega t \pm \frac{\pi}{3}\right)$

$x_1 = A$ at $t = 2.0$ s
$$A = A\cos(2\omega)$$
$$\frac{A}{A} = \cos(2\omega)$$
$$1 = \cos(2\omega)$$
$$2\omega = 2k\pi$$
$$\omega = k\pi \text{ where } k \text{ is an integer}$$

$x_2(t) = A\cos\left(k\pi t \pm \frac{\pi}{3}\right)$

The second oscillator reaches $x = A$ at the following time:

$$A = A\cos\left(k\pi t \pm \frac{\pi}{3}\right)$$

$$1 = \cos\left(k\pi t \pm \frac{\pi}{3}\right)$$

$$k\pi t \pm \frac{\pi}{3} = n2\pi \text{ where } n \text{ is an integer}$$

$$t = \frac{n2\pi \mp \frac{\pi}{3}}{k\pi}$$

$$t = \frac{2n \mp \frac{1}{3}}{k}$$

The smallest values of t are $t = 0.3$ s and $t = 1.7$ s (for $n = k = 0$); which is correct depends on whether the second oscillator leads or lags the first oscillator.

27. (a) The velocity of the simple harmonic oscillator is given by the equation

$$v(t) = -(10 \text{ cm/s})\sin(4t + \pi/2)$$

Comparing this equation to the standard equation for the velocity of a simple harmonic oscillator, $v(t) = -(\omega A)\sin(\omega t + \phi)$, we have

$\omega = 4$ rad/s

$$T = \frac{2\pi}{\omega} = \frac{\pi}{2} \text{ s}$$

Since $\omega A = 10$ cm/s, $A = (10 \text{ cm/s})/4 \text{ s}^{-1} = 2.5$ cm and $\phi = \pi/2$ rad.

(b) Because we know A, ω, and ϕ, we can write the equation for the displacement of this oscillator as a function of time:

$$x(t) = (2.5 \text{ cm})\cos(4t + \pi/2)$$

The particle passes through the equilibrium position when $x(t) = 0$. This occurs when

$$\cos(4t + \pi/2) = 0$$

$$4t + \frac{\pi}{2} = \left(m + \frac{1}{2}\right)\pi$$

$$t = m\frac{\pi}{4}, \quad m = 0, 1, 2, \ldots$$

Therefore, the first time the particle passes through the equilibrium position after $t = 0$ is at

$$t = \frac{\pi}{4} \text{ s}$$

(c) Since $x(t) = (2.5 \text{ cm}) \cos(4t + \pi/2)$, the maximum displacement is 2.5 cm. This occurs when

$$\cos(4t + \pi/2) = 1$$

$$4t + \frac{\pi}{2} = m(2\pi), \quad m = 0, 1, 2, \ldots$$

$$t = m\frac{\pi}{2} - \frac{\pi}{8}, \quad m = 0, 1, 2, \ldots$$

For $m = 0$, t is negative. The first time after $t = 0$ when the particle is at $x = +2.5$ cm occurs for $m = 1$ at

$$t = \frac{\pi}{2} \text{ s} - \frac{\pi}{8} \text{ s} = \frac{3\pi}{8} \text{ s}$$

(d) The total distance covered in one complete oscillation is $4A = 10$ cm.

(e) $a_{max} = \omega^2 A = (4 \text{ rad/s})^2 (2.5 \text{ cm}) = 40.0 \text{ cm/s}^2$

(f) $a(t) = -(40.0 \text{ cm/s}^2) \cos(4t + \pi/2)$

29. No. Applying the given information to the position function $x(t) = A\cos(\omega t + \phi)$, we get

$$x(0) = A\cos(\phi) \tag{1}$$

and

$$x\left(\frac{T}{2}\right) = A\cos\left(\omega\frac{T}{2} + \phi\right)$$

$$= A\cos\left(\frac{2\pi}{T} \times \frac{T}{2} + \phi\right)$$

$$= A\cos(\pi + \phi)$$

$$= -A\cos(\phi) \tag{2}$$

Equations (1) and (2) are not independent, so there is no way to use only these two equations to solve for the values A and ϕ, given the values of $x(0)$ and $x(T/2)$.

31. (a) The lower the angular frequency (the coefficient of t), the higher the period; therefore, the periods, from high to low, are $3 > 2 > 4 > 1 > 5$.

(b) The maximum speed is ωA, so the maximum speeds from high to low are $5 > 1 > 4 > 2 > 3$.

(c) The phase angles, from high to low, are $2 > 4 > 1 > 3 = 5$.

(d) The maximum acceleration is $\omega^2 A$, so the maximum accelerations from high to low are $5 > 1 > 4 > 2 > 3$.

33. Considering only the rotational motion of Earth, the period of a point on the equator is 1 day. The angular speed is 1 revolution per day, which is

$$\frac{2\pi \text{ rad}}{\text{day}} = \frac{2\pi \text{ rad}}{24 \times 3600 \text{ s}} = \frac{\pi \text{ rad}}{43\,200 \text{ s}} = 7.27 \times 10^{-5} \text{ rad/s}$$

$T = 86\,400$ s

35. Because the period is

$$T = 2\pi\sqrt{\frac{m}{k}}$$

the rank of the periods, from small to large, is D < A < B < E < F < C.

37. (a) $\omega = \sqrt{\dfrac{k}{m}} = \sqrt{\dfrac{100. \text{ N/m}}{0.10 \text{ kg}}} = \sqrt{1.0 \times 10^3 \text{ s}^{-2}} = 3.2 \times 10^1$ rad/s

$$T = \frac{2\pi}{\omega} = \frac{2\pi}{\sqrt{1.0 \times 10^3 \text{ s}^{-2}}} = 0.20 \text{ s}$$

(b) $v_{max} = \omega A = \sqrt{1.0 \times 10^3 \text{ s}^{-2}} \times 0.20 \text{ m} = 6.3$ m/s

$a_{max} = \omega^2 A = (\sqrt{1.0 \times 10^3 \text{ s}^{-2}})^2 (0.20 \text{ m}) = 2.0 \times 10^2$ m/s²

(c) $E = \dfrac{1}{2} kA^2 = \dfrac{1}{2}(100. \text{ N/m})(0.20 \text{ m})^2 = 2.0$ J

39. (a) Both oscillators will have the same amplitude because each is displaced by the same distance from its equilibrium position and then released from rest. The harmonic motion will be centred at the freely hanging position of each mass.

(b) Since each oscillator has identical springs (the same spring constant) and has the same amplitude, the total energy of the oscillators will be the same.

(c) The maximum speed of a mass–spring oscillator is given by

$$v_{max} = \omega A = \sqrt{\frac{k}{m}}\, A$$

For the oscillator with mass m, $v_{max} = \sqrt{\dfrac{k}{m}}\, A$.

For the oscillator with mass $2m$, $v_{max} = \sqrt{\dfrac{k}{2m}}\, A$.

Therefore, the maximum speed of the oscillator with mass $2m$ is a factor of $1/\sqrt{2}$ smaller than that of the oscillator with mass m.

41. (a) The position and velocity of the oscillator at $t = 0$ are

$x(0) = 0.20$ m

$v(0) = 1.0$ m/s

and $\omega = 2\pi/T = 2\pi/(2 \text{ s}) = \pi$ rad/s.

We start with general equations for the position and velocity of a simple harmonic oscillator:

$x(t) = A\cos(\omega t + \phi)$

$v(t) = -(\omega A)\sin(\omega t + \phi)$

At $t = 0$ the above equations reduce to

$x(0) = A\cos(\phi)$ \hfill (1)

$v(0) = -(\omega A)\sin(\phi)$

$\dfrac{v(0)}{\omega} = -A\sin(\phi)$ \hfill (2)

Squaring and adding (1) and (2), we get

$x(0)^2 + \dfrac{v(0)^2}{\omega^2} = A^2\cos^2(\phi) + A^2\sin^2(\phi)$

$x(0)^2 + \dfrac{v(0)^2}{\omega^2} = A^2[\cos^2(\phi) + \sin^2(\phi)]$

$x(0)^2 + \dfrac{v(0)^2}{\omega^2} = A^2$

$A = \sqrt{x(0)^2 + \dfrac{v(0)^2}{\omega^2}}$ \hfill (3)

and dividing (2) by (1), we get

$\tan(\phi) = -\dfrac{v(0)}{\omega x(0)} \Rightarrow \phi = \tan^{-1}\left(-\dfrac{v(0)}{\omega x(0)}\right)$ \hfill (4)

Let us take the velocity of the oscillator to be positive when the mass is moving upward and negative when the mass is moving downward. Inserting values of $x(0)$ and $v(0)$, we get

$A = \sqrt{(0.20 \text{ m})^2 + \dfrac{(1.0 \text{ m/s})^2}{(\pi \text{ rad/s})^2}} = 0.38$ m

$\phi = \tan^{-1}\left(-\dfrac{1.0 \text{ m/s}}{(\pi \text{ rad/s})(0.20 \text{ m})}\right) = -1.0$ rad

(b) When the mass passes through the equilibrium position, it has maximum speed

$$v_{max} = \omega A = (\pi \text{ rad/s})(0.38 \text{ m}) = 1.2 \text{ m/s}$$

(c) From (3) we observe that the amplitude depends upon the square of it velocity at $t = 0$. Hence the direction of motion at $t = 0$ does not affect the amplitude of oscillations. However, the phase constant does depend upon $v(0)$; therefore, if the sign of the velocity changes, the sign of the phase constant will also change. If the initial velocity is directed downward, then $v(0) = -1.0$ m/s, so

$$\phi = \tan^{-1}\left(-\frac{-1.0 \text{ m/s}}{(\pi \text{ rad/s})(0.20 \text{ m})}\right) = +1.0 \text{ rad}$$

43. (a) By symmetry, the block remains in contact with the spring for one half of its cycle of oscillation. The period of oscillation is

$$T = 2\pi\sqrt{\frac{m}{k}}$$
$$= 2\pi\sqrt{\frac{0.5 \text{ kg}}{50 \text{ N/m}}}$$
$$= \frac{\pi}{5} \text{ s}$$

Thus, the block remains in contact with the spring for $\pi/5$ s $\div$ 2 $= \pi/10$ s $= 0.31$ s.

(b) The initial speed of the block played no role in the calculation of part (a), so changes in the initial speed of the block do not change the time of contact. (The amplitude of the oscillation depends on the initial speed of the block, but not the period of the oscillation.)

45. Since the distance between the two turning points of a simple harmonic motion is $2A$,

$$2A = 6.0 \text{ cm} \Rightarrow A = 3.0 \text{ cm}$$

This means that the equilibrium position, with the mass hanging motionlessly with the spring, is 3.0 cm below the unstretched position of the spring. Thus,

$$k\Delta L = mg$$
$$\frac{m}{k} = \frac{\Delta L}{g}$$

The period of oscillation is

$$T = 2\pi\sqrt{\frac{m}{k}} = 2\pi\sqrt{\frac{\Delta L}{g}}$$
$$= 2\pi\sqrt{\frac{3.0 \times 10^{-2} \text{ m}}{9.81 \text{ m/s}^2}} = 0.35 \text{ s}$$

47. Suppose $x_1 = A\cos(\omega t)$ and $x_2 = B\cos(\omega t)$. Then

$$a_1 = -\frac{k}{m_1}x_1 + \frac{k}{m_1}x_2$$

$$a_2 = \frac{k}{m_2}x_1 - \frac{k}{m_2}x_2$$

Substituting for the accelerations, we obtain

$$-\omega^2 A = -\frac{k}{m_1}A + \frac{k}{m_1}B$$

$$0 = \left(\omega^2 - \frac{k}{m_1}\right)A + \frac{k}{m_1}B \qquad (1)$$

$$-\omega^2 B = \frac{k}{m_2}A - \frac{k}{m_2}B$$

$$0 = \frac{k}{m_2}A + \left(\omega^2 - \frac{k}{m_2}\right)B \qquad (2)$$

Since amplitudes A and B must be non-zero, (1) and (2) are satisfied provided that

$$\left(\omega^2 - \frac{k}{m_1}\right)\left(\omega^2 - \frac{k}{m_2}\right) - \frac{k^2}{m_1 m_2} = 0$$

$$\omega^4 - k\left(\frac{1}{m_1} + \frac{1}{m_2}\right)\omega^2 = 0$$

$$\omega^2 = k\left(\frac{1}{m_1} + \frac{1}{m_2}\right)$$

$$\omega = \sqrt{k\left(\frac{1}{m_1} + \frac{1}{m_2}\right)}$$

$$f = \frac{\omega}{2\pi} = \frac{1}{2\pi}\sqrt{k\left(\frac{1}{m_1} + \frac{1}{m_2}\right)}$$

49. (a) Suppose the block is displaced from its equilibrium position by a small distance x toward the right. Then spring 2 is compressed by a length x and spring 1 is stretched by a length x. The directions of the forces exerted by springs 1 and 2 on the mass pushed the mass toward its equilibrium position. The net force, F, exerted on the mass is therefore

$$F = -k_1 x - k_2 x$$
$$= -(k_1 + k_2)x$$

which implies simple harmonic motion with an effective spring constant of $k_1 + k_2$.

(b) $\omega^2 = \dfrac{k_1 + k_2}{m} = \dfrac{k_1}{m} + \dfrac{k_2}{m} = \omega_1^2 + \omega_2^2$

Since $\omega = 2\pi f$,

$f^2 = f_1^2 + f_2^2$

(c) $U_1 = \dfrac{1}{2}k_1 x^2 \quad U_2 = \dfrac{1}{2}k_2 x^2 \;\Rightarrow\; \dfrac{U_1}{U_2} = \dfrac{k_1}{k_2}$

(d) No. Each spring stretches or compresses by x.

51. (a) $|F_{max}| = kA \to A = \dfrac{|F_{max}|}{k} = \dfrac{3.5 \text{ N}}{200. \text{ N/m}} = 0.0175 \text{ m} = 1.8 \text{ cm}$

(b) $E = \dfrac{1}{2}kA^2 = \dfrac{1}{2}(200. \text{ N/m})(0.0175 \text{ m})^2 = 0.031 \text{ J}$

(c) $\dfrac{1}{2}mv_{max}^2 = \dfrac{1}{2}kA^2$

$v_{max} = \sqrt{\dfrac{k}{m}A^2} = \sqrt{\dfrac{k}{m}}A = \sqrt{\dfrac{200. \text{ N/m}}{2 \text{ kg}}}(0.0175 \text{ m}) = 0.18 \text{ m/s} = 1.8 \times 10^1 \text{ cm/s}$

53. $E = K + U$

When $U = K$, $E = 2U$, or $U = E/2$. Thus,

$\dfrac{1}{2}kx^2 = \dfrac{1}{2}\left(\dfrac{1}{2}kA^2\right)$

$x^2 = \dfrac{1}{2}A^2$

$x = \pm\dfrac{1}{\sqrt{2}}A$

55. (a) For this problem we are given that

$k = 10 \text{ N/m}$

$m = 200 \text{ g} = 0.2 \text{ kg}$

Also, at $t = 0$, $x(0) = -10$ cm and $v(0) = 0$ cm/s. Therefore, the amplitude is (see the solution to problem 41)

$A = \sqrt{x(0)^2 + \dfrac{v(0)^2}{\omega^2}} = 10 \text{ cm}$

The phase constant is (see the solution to problem 41),

$$\phi = \tan^{-1}\left(\frac{v(0)}{\omega x(0)}\right) = \tan^{-1}(0) = 0 \text{ rad}$$

The angular frequency, ω, is

$$\omega = \sqrt{\frac{k}{m}} = \sqrt{\frac{10 \text{ N/m}}{0.200 \text{ kg}}} = 7 \text{ rad/s}$$

(b) The total energy of the oscillator is

$$E = \frac{1}{2}kA^2 = \frac{1}{2}(10 \text{ N/m})(0.10 \text{ m})^2 = 5\times10^{-2} \text{ J}$$

57. To determine k, note that when the spring is in equilibrium,
$mg = k\Delta x$

$$k = \frac{mg}{\Delta x} = \frac{(2.0 \text{ kg})(9.81 \text{ m/s}^2)}{(0.100 \text{ m})} = 1.96\times10^2 \text{ N/m}$$

(a) $A = 5.0$ cm $= 0.050$ m

$$f = \frac{1}{2\pi}\sqrt{\frac{k}{m}} = \frac{1}{2\pi}\sqrt{\frac{1.96\times10^2 \text{ N/m}}{2.0 \text{ kg}}} = 1.6 \text{ s}^{-1}$$

(b) $K = \frac{1}{2}kA^2 - \frac{1}{2}kx^2$

$$= \frac{1}{2}k(A^2 - x^2)$$

$$= (98.1 \text{ N/m})[(0.05 \text{ m})^2 - (0.03 \text{ m})^2]$$

$$= 0.16 \text{ J}$$

(c) $A = 0.050$ m and $\omega = 2\pi f = 9.9$ rad/s. At $t = 0$, the displacement from the equilibrium position is -0.050 m, so $\phi = \pi$ rad. Therefore, the equation for this simple harmonic motion is given by

$y(t) = (0.050 \text{ m})\cos(9.9t + \pi)$

Using $\cos(\theta + \pi) = -\cos(\theta)$,

$y(t) = -(0.050 \text{ m})\cos(9.9t)$

(d) Half as much: 5.0 cm, since each spring will stretch by an equal amount.

59. (a) $T = 2\pi\sqrt{\dfrac{m}{k}} \to k = \dfrac{4\pi^2 m}{T^2} = \dfrac{4\pi^2 (1.0 \text{ kg})}{(2.0 \text{ s})^2} = 9.9$ N/m

(b) $\dfrac{1}{2}kA^2 = \dfrac{1}{2}(9.9 \text{ N/m})(0.05 \text{ m})^2 = 0.012$ J

(c) $U = \dfrac{1}{2}k(0.03 \text{ m})^2 \Rightarrow K = \dfrac{1}{2}kA^2 - \dfrac{1}{2}k(0.03 \text{ m})^2$

$\dfrac{K}{E} = \dfrac{\dfrac{1}{2}kA^2 - \dfrac{1}{2}k(0.03 \text{ m})^2}{\dfrac{1}{2}kA^2} = 1 - \dfrac{(0.03 \text{ m})^2}{(0.05 \text{ m})^2} = 1 - \left(\dfrac{3}{5}\right)^2 = 1 - \dfrac{9}{25} = 64\%$

(d) $x = (0.050 \text{ m})\cos(\pi t)$ because $\omega = \sqrt{\dfrac{k}{m}} = \dfrac{2\pi}{T} = \dfrac{2\pi}{2.0 \text{ s}} = \pi \text{ s}^{-1}$.

61. (a) $T = 2\pi\sqrt{\dfrac{L}{g}} \Rightarrow g = \dfrac{4\pi^2 L}{T^2} = \dfrac{4\pi^2 (1.00 \text{ m})}{(2.00 \text{ s})^2} = \pi^2 \text{ m/s}^2 = 9.87 \text{ m/s}^2$

(b) Increase; $T \propto 1/\sqrt{g}$, so because the value of g is less at the top of Mount Everest, the value of T will be greater.

(c) On the Moon, $g = 1.62$ m/s^2. The period on the Moon is

$T = 2\pi\sqrt{\dfrac{L}{g}} = 2\pi\sqrt{\dfrac{1.00 \text{ m}}{1.62 \text{ m/s}^2}} = 4.94$ s

(d) The International Space Station is in free fall, so the pendulum will not oscillate there.

63. $h = 1 - \cos(15°)$
    $= 3.4$ cm

(a) $E = mgh = (0.30 \text{ kg})(9.81 \text{ m/s}^2)(0.034 \text{ m})$
    $= 0.10$ J

(b) $\dfrac{1}{2}mv^2 = mgh$
    $v = \sqrt{2gh}$
    $v = \sqrt{2(9.81 \text{ m/s}^2)(0.034 \text{ m})}$
    $v = 0.82$ m/s

65. (a) $T = 2\pi\sqrt{\dfrac{L}{g}} = \dfrac{2\pi}{\sqrt{g}}L^{1/2}$

$\dfrac{dT}{dL} = \dfrac{2\pi}{\sqrt{g}}\left(\dfrac{1}{2}L^{-1/2}\right)$

$dT = \dfrac{2\pi}{\sqrt{g}}\dfrac{1}{\sqrt{L}}\left(\dfrac{1}{2}dL\right)$

$\dfrac{dT}{T} = \dfrac{2\pi g^{-1/2}L^{-1/2}2^{-1}dL}{2\pi L^{1/2}g^{-1/2}} = \dfrac{dL}{2L}$

$dT = \left(\dfrac{T}{2L}\right)dL$

(b) $\dfrac{dT}{T} = \dfrac{dL}{2L} \Rightarrow \dfrac{dL}{L} = 2\dfrac{dT}{T}$

$\dfrac{dT}{T} = +\dfrac{5\text{ s}}{1\text{ h}} = +\dfrac{5\text{ s}}{3600\text{ s}} = +\dfrac{5}{3600}$

The positive sign is because we wish to increase the period.

$\dfrac{dL}{L} = 2\dfrac{dT}{T} = +\dfrac{10}{3600} = +\dfrac{1}{360} = 0.28\%$

Thus, the length of the pendulum must be increased by 0.28%.

67. $I_p = \dfrac{2}{5}mR^2 + mL^2$

$T = 2\pi\sqrt{\dfrac{I_p}{mgL}}$

$= 2\pi\sqrt{\dfrac{\dfrac{2}{5}mR^2 + mL^2}{mgL}}$

$= 2\pi\sqrt{\dfrac{2R^2 + 5L^2}{5gL}}$

$= 2\pi\sqrt{\dfrac{L}{g} + \dfrac{2R^2}{5gL}}$

$= 2\pi\sqrt{\dfrac{L}{g}\left[1 + \dfrac{2R^2}{5L^2}\right]}$

$$= 2\pi\sqrt{\frac{L}{g}}\sqrt{1+\frac{2R^2}{5L^2}}$$

$$= T_0\sqrt{1+\frac{2R^2}{5L^2}}$$

69. (a) Reading from the graph, the amplitude is 0.6 m and the period is 3.0 s.

    (b) If we use a cosine function to model the position function, then the phase constant is
    $$\frac{0.75 \text{ s}}{3 \text{ s}} \times 2\pi = \frac{\pi}{2}.$$

    (c) The phase at $t = 1$ s is
    $$\frac{2\pi}{3}\text{ s}^{-1}(1 \text{ s}) + \frac{\pi}{2} = \frac{7\pi}{6}$$

    The phase at $t = 2$ s is
    $$\frac{2\pi}{3}\text{ s}^{-1}(2 \text{ s}) + \frac{\pi}{2} = \frac{11\pi}{6}$$

    (d) $x(t) = (0.6 \text{ m})\cos\left(\frac{2\pi}{3}t + \frac{\pi}{2}\right)$

    (e) Differentiating the formula for the position function, we obtain
    $$v(t) = -(0.4\pi \text{ m/s})\sin\left(\frac{2\pi}{3}t + \frac{\pi}{2}\right)$$

    Thus,
    $$v(1.0) = -(0.4\pi \text{ m/s})\sin\left(\frac{2\pi}{3} + \frac{\pi}{2}\right)$$
    $$= -(0.4\pi \text{ m/s})\sin\left(\frac{7\pi}{6}\right)$$
    $$= 0.63 \text{ m/s}$$

    (f) Differentiating the formula for the velocity function, we obtain
    $$a(t) = -\left(\frac{0.8\pi^2}{3} \text{ m/s}^2\right)\cos\left(\frac{2\pi}{3}t + \frac{\pi}{2}\right)$$

Thus,

$$a(2) = -\left(\frac{0.8\pi^2}{3} \text{ m/s}^2\right)\cos\left(\frac{4\pi}{3} + \frac{\pi}{2}\right)$$

$$= -\left(\frac{0.8\pi^2}{3} \text{ m/s}^2\right)\cos\left(\frac{11\pi}{6}\right)$$

$$= -2.3 \text{ m/s}^2$$

71. (a) Reading from the graph, the amplitude of the oscillation is 0.2 m, and the period is 3 s.

    (b) The equilibrium position is 1 m.

    (c) Reading from the graph, the phase constant is $\frac{0.5 \text{ s}}{3 \text{ s}} \times 2\pi = \frac{\pi}{3}$.

    (d) $x(t) = A\cos(\omega t + \phi) + 1 = (0.2 \text{ m})\cos\left(\frac{2\pi}{3}t + \frac{\pi}{3}\right) + 1$

    (e) $x(t) = (0.2 \text{ m})\cos\left(\frac{2\pi}{3}t + \frac{\pi}{3} \pm \frac{\pi}{2}\right) + 1$

    (f) $x(t) = (0.2 \text{ m})\cos\left(\frac{2\pi}{3}t + \frac{\pi}{3}\right) + 1$

    $$v(t) = \left(-\frac{0.4\pi}{3} \text{ m/s}\right)\sin\left(\frac{2\pi}{3}t + \frac{\pi}{3}\right)$$

    $$a(t) = \left(-\frac{0.8\pi^2}{9} \text{ m/s}^2\right)\cos\left(\frac{2\pi}{3}t + \frac{\pi}{3}\right)$$

    $$a(1) = \left(-\frac{0.8\pi^2}{9} \text{ m/s}^2\right)\cos\left(\frac{2\pi}{3}(1) + \frac{\pi}{3}\right)$$

    $$a(1) = \left(-\frac{0.8\pi^2}{9} \text{ m/s}^2\right)\cos(\pi)$$

    $$a(1) = \left(-\frac{0.8\pi^2}{9} \text{ m/s}^2\right)(-1)$$

    $$a(1) = 0.88 \text{ m/s}^2$$

73. (a) 3, 1, 2

    (b) Because the springs are identical, the total energies are in the same rank order as the amplitudes; therefore, the order of total energies is 1, 3, 2.

    (c) 3, 2, 1

    (d) 1, 3, 2

75. (a) $e^{-bt/m} = 0.95$

$$-\frac{bT}{m} = \ln(0.95)$$

$$b = -\frac{m}{T}\ln(0.95) = -\frac{0.25 \text{ kg}}{2.0 \text{ s}}\ln(0.95) = 0.0064 \text{ kg/s}$$

(b) $e^{-bt/(2m)} = \frac{0.20 \text{ m}}{0.30 \text{ m}} = \frac{2}{3}$

$$e^{bt/(2m)} = \frac{3}{2}$$

$$\frac{bt}{2m} = \ln\left(\frac{3}{2}\right)$$

$$t = \frac{2m}{b}\ln(1.5)$$

$$\frac{t}{T} = \frac{2m}{bT}\ln(1.5)$$

$$\frac{t}{T} = \frac{2(0.25 \text{ kg})}{(0.0064 \text{ kg/s})(2 \text{ s})}\ln(1.5)$$

$$\frac{t}{T} = 15.8$$

(c) No, because the amplitude decreases according to a decreasing exponential function, not a linear function.

77. For A: $10e^{-50(0.01)} = 10e^{-0.5} = 6.1$ cm

For B: $20e^{-50(0.02)} = 20e^{-1} = 7.4$ cm

For C: $30e^{-50(0.03)} = 30e^{-1.5} = 6.7$ cm

A, C, B

79. The energy of a damped harmonic oscillator changes with time, given by

$$E(t) = E(0)e^{-t/\tau}$$

where $E(0)$ is the starting energy at time $t = 0$ and τ is the decay constant. After n periods, the energy of the oscillator is

$$E(nT) = E(0)e^{-nT/\tau}$$

Since the Q-value Q is given by $Q = 2\pi\frac{\tau}{T}$, we can write the above equation as

$$E(nT) = E(0)e^{-n(2\pi/Q)}$$

The energy of the oscillator after one oscillation ($n = 1$) is

$$E(T) = E(0)e^{-(2\pi/Q)}$$

The fraction of initial energy lost in one cycle, ΔE, is

$$\Delta E = \frac{E(0) - E(T)}{E(0)} = 1 - e^{-(2\pi/Q)}$$

It is given that 0.5% of the energy is lost in one cycle. Therefore,

$$\Delta E = \frac{0.5}{100} = 0.005 = 1 - e^{-(2\pi/Q)}$$

$$e^{-(2\pi/Q)} = 0.995$$

$$Q = -\frac{2\pi}{\ln(0.995)} \approx 1200$$

81. (a) $T = 2\pi\sqrt{\dfrac{L}{g}} = 2\pi\sqrt{\dfrac{15.0 \text{ m}}{9.81 \text{ m/s}^2}} = 7.77 \text{ s}$

    In one week, the number of oscillations is $\dfrac{7 \times 24 \times 3600 \text{ s}}{7.77 \text{ s}} = 77\,800$.

    (b) $e^{-b(5)/(2m)} = 0.8$

    $e^{-b/(2m)} = (0.8)^{1/5}$

    When the amplitude decreases from 100.0 cm to 20.0 cm,

    $$e^{-bt/(2m)} = \frac{20.0}{100.0} = 0.200$$

    $$(e^{-b/(2m)})^t = 0.200$$

    $$(0.8)^{t/5} = 0.200$$

    $$\frac{t}{5}\ln(0.8) = \ln(0.2)$$

    $$t = \frac{5\ln(0.2)}{\ln(0.8)}$$

    $$t = 36.1 \text{ days}$$

83. Suppose we start the oscillation by attaching a mass M to the upstretched spring and allowing it to fall freely. The amplitude and period of the oscillations are given by

    $$A = \frac{Mg}{k}, \quad T = 2\pi\sqrt{\frac{M}{k}}$$

When the mass is as its highest point (and hence stationary), we place a small mass m on the bigger mass M. Now a mass $(M + m)$ is attached to the same spring and is allowed to drop freely. The new amplitude and period are obtained by replacing M by $(M + m)$ in the above equation:

$$A' = \frac{(M+m)g}{k}, \quad T' = 2\pi\sqrt{\frac{(M+m)}{k}}$$

Since the energy of a simple harmonic oscillator is proportional to the square of its amplitude the energy will increase. Therefore,

(a) $A' > A$

(b) $T' > T$

(c) Energy increases.

85. $F = -\dfrac{GM_x m}{x^2}$, where $M_x = \rho V_x = \rho\left(\dfrac{4}{3}\pi x^3\right)$, $\rho = \dfrac{M}{V} = \dfrac{M}{\frac{4}{3}\pi R^3}$

$$M_x = \rho V_x = \frac{M}{\frac{4}{3}\pi R^3}\left(\frac{4}{3}\pi x^3\right) = \left(\frac{M}{R^3}\right)x^3$$

$$F = -\frac{GM_x m}{x^2}$$

$$F = -\frac{Gm}{x^2}\left(\frac{M}{R^3}x\right)^3$$

$$F = -\left(\frac{GMm}{R^3}\right)x$$

This is simple harmonic motion, with the effective spring constant being $k = \dfrac{Gmm}{R^3}$, where M = mass of Earth and R = radius of Earth. Thus,

$$\omega^2 = \frac{k}{m} = \frac{GM}{R^3} \Rightarrow T = \frac{2\pi}{\omega} = 2\pi\sqrt{\frac{R^3}{GM}}$$

The time to reach from one pole to the other (one half cycle) is

$$\frac{T}{2} = \pi\sqrt{\frac{R^3}{GM}} = \pi\sqrt{\frac{(6.371\times 10^6 \text{ m})^3}{(6.674\times 10^{-11} \text{ m}^3\text{ kg}^{-1}\text{ s}^{-2})(5.972\times 10^{24} \text{ kg})}} = 42.2 \text{ min}$$

which is independent of m.

87. Let ρ be the density of the liquid. When the object is floating in equilibrium, the weight of the liquid displaced by the object is equal to the weight of the object. Therefore,

$$m_{object}g = V_0\rho g$$

Suppose the object is now pushed farther into the liquid by a depth x. Then the additional buoyant force exerted on the object is equal to the weight of the additional liquid displaced:

$$F_B = (Ax)\rho g = (\rho A g)x$$

Notice that the magnitude of the additional buoyant force is proportional to x and the direction of the force is such that it pushes the object back to the equilibrium floating position. The object will therefore execute simple harmonic motion with effective spring constant $\rho A g$. The period of oscillation is

$$T = 2\pi\sqrt{\frac{m_{object}}{\rho A g}} = 2\pi\sqrt{\frac{\rho V_0}{\rho A g}} = 2\pi\sqrt{\frac{V_0}{gA}}$$

89. $U = \frac{1}{2}kx^2$

    $K = \frac{1}{2}mv^2$

    $E = \frac{1}{2}kA^2$

    $x(t) = A\cos(\omega t) = A\cos\left(2\pi\frac{t}{T}\right)$

    $v(t) = -\omega A \sin(\omega t)$

    (a) When $K = U$, $U = E/2$, so

    $$\frac{1}{2}kx^2 = \frac{1}{2}\left(\frac{1}{2}kA^2\right)$$

    $$x^2 = \frac{1}{2}A^2$$

    $$x = \pm\frac{1}{\sqrt{2}}A$$

    $$A\cos\left(2\pi\frac{t}{T}\right) = \pm\frac{1}{\sqrt{2}}A$$

    $$\cos\left(2\pi\frac{t}{T}\right) = \pm\frac{1}{\sqrt{2}}$$

    $$2\pi\frac{t}{T} = \pm\frac{\pi}{4} + 2n\pi$$

    $$t = \frac{T}{2}\left(2n \pm \frac{1}{4}\right), \text{ where } n \text{ is an integer}$$

(b) When $K = 2U$, we have

$$K + U = E$$
$$2U + U = E$$
$$3U = E$$
$$U = \frac{1}{3}E$$
$$\frac{1}{2}kx^2 = \frac{1}{3}\left(\frac{1}{2}kA^2\right)$$
$$x^2 = \frac{1}{3}A^2$$
$$x = \pm\frac{1}{\sqrt{3}}A$$

$$A\cos\left(2\pi\frac{t}{T}\right) = \pm\frac{1}{\sqrt{3}}A$$
$$2\pi\frac{t}{T} = \pm\cos^{-1}\left(\frac{1}{\sqrt{3}}\right) + 2n\pi, \ n \text{ is an integer}$$
$$t = \frac{T}{2}\left[2n \pm \frac{1}{\pi}\cos^{-1}\left(\frac{1}{\sqrt{3}}\right)\right]$$

Chapter —WAVES

In most solutions, we omit the units until the final answer to make calculations easier.

1. (a) False. For a transverse wave, the particles of the medium vibrate perpendicular to the direction of the wave. For a longitudinal wave, the particles move back and forth in the direction of the wave.

    (b) False. The wave speed of a mechanical wave in a medium depends on the elastic and inertial properties of the medium.

    (c) True. The speed of a wave in a medium can vary with wavelength; this is what causes the dispersion of sunlight into rainbow colours upon refraction in a prism. However, if we ignore dispersion as we have done in this chapter, then the wave speed is independent of the wavelength of the waves.

    (d) False. Mechanical waves depend on the movement of particles of the medium through which they pass.

    (e) False. Since the speed of the wave depends on the properties of the medium, the speed will be constant in a uniform medium.

    (f) False. A travelling wave always carries energy.

    (g) False. The maximum amplitude cannot be greater than the sum of the amplitudes of individual waves.

    (h) False. To produce a standing wave, the constituent waves must be travelling in opposite directions.

3. The light wave has the longer wavelength in this case, although it would be more accurate to call it an electromagnetic wave, because at this wavelength it is in the radio wave part of the electromagnetic spectrum.

    $$v_s = \lambda_s f_s$$
    $$\lambda_s = \frac{v_s}{f_s} = \frac{340 \text{ m/s}}{20\,000 \text{ s}^{-1}} = 1.70 \times 10^{-2} \text{ m}$$
    $$v_\ell = \lambda_\ell f_\ell$$
    $$\lambda_\ell = \frac{v_\ell}{f_\ell} = \frac{3 \times 10^8 \text{ m/s}}{20\,000 \text{ s}^{-1}} = 1.5 \times 10^4 \text{ m}$$

5. No. The speed of individual particles in the medium is not the same as the wave speed through the medium.

7. The wave propagates outward from the point where the stone hits the water surface. As the wave front expands, the energy per unit length decreases, and therefore the amplitude decreases.

9. The answer is (c) for a mechanical wave, as shown below (see Section 14-8):

$$\Delta E = \Delta K + \Delta U$$
$$\Delta E_1 = (\mu \Delta x)\omega_1^2 A^2 \cos^2(kx - \omega_1 t)$$
$$\Delta E_2 = (\mu \Delta x)\omega_2^2 A^2 \cos^2(kx - \omega_2 t) = (\mu \Delta x)(2\omega_1)^2 A^2 \cos^2(kx - 2\omega_1 t)$$
$$= 4(\mu \Delta x)\omega_1^2 A^2 \cos^2(kx - 2\omega_1 t) = 4\Delta E_1$$

In words, the total energy of a mechanical wave is proportional to the square of the frequency, so if the frequency is doubled, the energy increases by a factor of four.

11. The two waves must have same frequency and amplitude to produce a standing wave. If the two waves have the same amplitude but different frequencies, interference will occur but standing waves will not be produced. If the two waves have the same frequency but different amplitudes, a pattern similar to a standing wave will be produced but without any nodes.

13. No; a difference in linear mass density could be offset by a difference in tension.

15. (a) $k = \dfrac{2\pi}{\lambda} = \dfrac{2\pi}{2.00 \text{ m}} = \pi \text{ m}^{-1}$

   (b) $\omega = 2\pi f = 2\pi(10 \text{ Hz}) = 63 \text{ rad/s}$

   (c) $v = \lambda f = 2.00 \text{ m}(10 \text{ s}^{-1}) = 20 \text{ m/s}$

17. (a) $\omega = kv = 2\pi f$

   $f = \dfrac{kv}{2\pi} = \dfrac{(6.00 \text{ rad/m})(150.0 \text{ m/s})}{2\pi} = 143 \text{ Hz}$

   (b) $T = \dfrac{1}{f} = \dfrac{1}{143 \text{ s}^{-1}} = 6.99 \times 10^{-3} \text{ s}$

   (c) $k = \dfrac{2\pi}{\lambda}$

   $\lambda = \dfrac{2\pi}{k} = \dfrac{2\pi}{6.00 \text{ rad/m}} = 1.05 \text{ m}$

19. $\lambda = \dfrac{v}{f}$

$\lambda_1 = \dfrac{v}{f_1} = \dfrac{340 \text{ m/s}}{20 \text{ s}^{-1}} = 17 \text{ m}$

$\lambda_2 = \dfrac{v}{f_2} = \dfrac{340 \text{ m/s}}{20\,000 \text{ s}^{-1}} = 17 \text{ mm}$

The range of wavelengths is from 17 mm to 17 m.

21. Wavelength range in the ocean:

$f_1 = 250 \text{ Hz}$
$f_2 = 150\,000 \text{ Hz}$
$v = 1500 \text{ m/s}$

$\lambda_1 = \dfrac{v}{f_1} = \dfrac{1500 \text{ m/s}}{250 \text{ s}^{-1}} = 6.00 \text{ m}$

$\lambda_2 = \dfrac{v}{f_2} = \dfrac{1500 \text{ m/s}}{150\,000 \text{ s}^{-1}} = 10. \text{ mm}$

Wavelength range in air:

$f_1 = 250 \text{ Hz}$
$f_2 = 150\,000 \text{ Hz}$
$v = 340 \text{ m/s}$

$\lambda_1 = \dfrac{v}{f_1} = \dfrac{340 \text{ m/s}}{250 \text{ s}^{-1}} = 1.4 \text{ m}$

$\lambda_2 = \dfrac{v}{f_2} = \dfrac{340 \text{ m/s}}{150\,000 \text{ s}^{-1}} = 2.3 \times 10^{-3} \text{ m}$

23. The pulse speed is 4.0 m/s, and the pulse is travelling in the direction of increasing x, so in the equation for $D(x,0)$ we make the following substitution:

$x \rightarrow x - 4.0t$

The equation for the pulse for any time is then given by

$D(x,t) = \dfrac{0.3}{(x - 4.0t)^2 + 1.2}$

At $x = 1.5$ m and $t = 2.0$ s, the displacement is given by

$$D(1.5, 2.0) = \frac{0.3 \text{ m}^3}{[(1.5 \text{ m}) - (4.0 \text{ m/s})(2.0 \text{ s})]^2 + 1.2 \text{ m}^2} = 6.9 \times 10^{-3} \text{ m}$$

The maximum displacement is when the quantity $x - vt$ is zero. The maximum displacement is 0.25 m.

The minimum displacement is zero:

$$\lim_{t \to \infty} D(x,t) = \frac{0.3}{(x-vt)^2 + 1.2} = 0$$

25. (a) $v = 3.0$ m/s and the pulse is travelling in the direction of the negative x-axis.

    (b) $D(2.0 \text{ m}, 3.0 \text{ s}) = \dfrac{-3.0 \text{ m}^3}{6.0 \text{ m}^2 + [(2.0 \text{ m}) + 3.0 \text{ m/s}(3.0 \text{ s})]^2} = -2.4$ cm

    (c) The maximum value for the displacement occurs when the value $x + 3.0t$ is zero, so $D = -0.5$ m.

    (d) $D(x,t) = \dfrac{-3.0 \text{ m}^3}{6.0 \text{ m}^2 + [x + (3.0 \text{ m/s})t]^2}$

    $u(x,t) = \dfrac{\partial}{\partial t} D(x,t) = -[(-3.0)(6.0 + (x + 3.0t)^2)^{-2}][2(3.0)(x + 3.0t)]$

    $u(2.0, 3.0) = [(3.0)[6.0 + (2.0 + 3.0(3.0))^2]^{-2}][2(3.0)(2.0 + 3.0(3.0))] = 1.2$ cm/s

    (e)
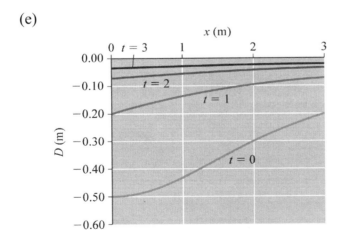

27. (a)

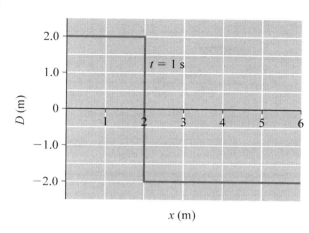

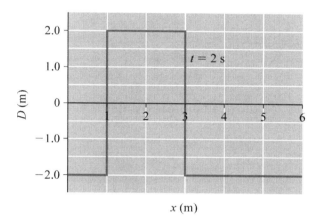

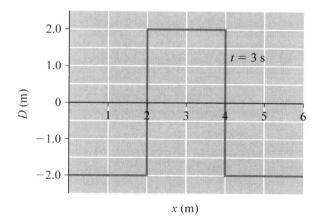

(b) Reading from the graph, the leading edge of the wave moves 1.0 m to the right every second, so the velocity of the pulse is 1.0 m/s to the right.

(c) For the given values of x and t, $|0.5-1.0| \leq 1$, so the displacement is 2 m.

(d) $D(x,t) = \begin{cases} +2 \text{ m} & \text{if } |x+t| \leq 1 \\ -2 \text{ m} & \text{if } |x+t| > 1 \end{cases}$

29. (a)

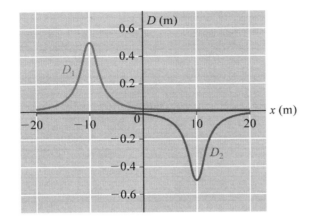

(b) The peak of each pulse occurs when the expression in parentheses is zero. At $t = 0$, the peak of the D_1 pulse is at $x = -10$ m, and the peak of the D_2 pulse is at $x = 10$ m.

(c) The condition corresponding to the cancelling of the pulses is

$$\frac{2}{(x-5t+10)^2+4} + \frac{-2}{(x+5t-10)^2+4} = 0$$
$$2[(x+5t-10)^2+4] - 2[(x-5t+10)^2+4] = 0$$
$$(x+5t-10)^2+4 - [(x-5t+10)^2+4] = 0$$
$$(x+5t-10)^2+4 - (x-5t+10)^2 - 4 = 0$$
$$x^2+10tx-20x+25t^2-100t+100 - (x^2-10tx+20x+25t^2-100t+100) = 0$$
$$20tx - 40x = 0$$
$$20x(t-2) = 0$$

Thus, the pulses cancel when $t = 2$ s.

(d) From the equation at the end of the calculation in part (c), when $x = 0$ the pulses cancel for all times.

31. First find the linear mass density of the wire:

$$\mu = \rho A = (7.86 \text{ g/cm}^3)[\pi(0.020 \text{ cm})^2] = 9.88 \times 10^{-3} \text{ g/cm} = 9.88 \times 10^{-4} \text{ kg/m}$$

Then use the formula relating velocity, tension, and linear mass density to calculate the tension in the wire:

$$v = \sqrt{\frac{T_s}{\mu}}$$

$$T_s = v^2 \mu = (160 \text{ m/s})^2 (9.88 \times 10^{-4} \text{ kg/m}) = 25 \frac{\text{kg} \cdot \text{m}}{\text{s}^2} = 25 \text{ N}$$

33. $\mu = \frac{M}{L} = \frac{0.10 \text{ kg}}{50.0 \text{ m}} = 2.0 \times 10^{-3} \text{ kg/m}$

$$v = \sqrt{\frac{T_s}{\mu}} = \sqrt{\frac{100.0 \text{ N}}{2.0 \times 10^{-3} \text{ kg/m}}} = 220 \text{ m/s}$$

35. First write the linear mass density of both strings in kg/m:

$\mu_A = 2.0 \text{ g/m} = 2.0 \times 10^{-3} \text{ kg/m}$

$\mu_B = 5.0 \text{ g/m} = 5.0 \times 10^{-3} \text{ kg/m}$

Then calculate the speed of each pulse using the same value for the tension since the strings are attached and the tension is constant:

$$v_A = \sqrt{\frac{T}{\mu_A}} = \sqrt{\frac{50.0 \text{ N}}{2.0 \times 10^{-3} \text{ kg/m}}} = 1.6 \times 10^2 \text{ m/s}$$

$$v_B = \sqrt{\frac{T}{\mu_B}} = \sqrt{\frac{50.0 \text{ N}}{5.0 \times 10^{-3} \text{ kg/m}}} = 1.0 \times 10^2 \text{ m/s}$$

Now calculate the time it takes for pulse A to travel the length of string A (10.0 m). Then calculate how far pulse B would travel in the same time. After this time, the pulses will be travelling on string B only and will therefore travel at the same speed:

$$t_A = \frac{d_A}{v_A} = \frac{10.0 \text{ m}}{1.6 \times 10^2 \text{ m/s}} = 6.2 \times 10^{-2} \text{ s}$$

$$d_B = v_B t_A = (1.0 \times 10^2 \text{ m/s})(6.2 \times 10^{-2} \text{ s}) = 6.2 \text{ m}$$

So the pulses are now 20.0 m − 6.2 m = 13.8 m apart, and they travel at the same speed in string B. Therefore, they meet halfway between their current positions, 6.2 m + 6.85 m = 13.0 m from the far end of string B.

37. (a) $[v] = (\text{m} \cdot \text{s}^{-2} \cdot \text{m})^{1/2} = \text{m/s}$

(b) $T = \frac{1}{f} = \frac{\lambda}{v} = \lambda \sqrt{\frac{2\pi}{g\lambda}} = \sqrt{\frac{2\pi\lambda}{g}}$

(c) From the result of part (b),

$$T^2 = \frac{2\pi\lambda}{g}$$

Therefore,

$$\lambda = \frac{T^2 g}{2\pi}$$

Since $v = \frac{\lambda}{T}$, we have,

$$v = \frac{gT}{2\pi} = \frac{(9.81 \text{ m/s}^2)(12 \text{ s})}{2\pi} = 19 \text{ m/s}$$

39. All four waves move in the positive x-direction, because the argument of the sinusoidal function in each case is of the form $kx - \omega t$. Because the wave number k is the reciprocal of the wavelength, the rank of the waves in order of increasing wavelength is b < a = c < d. The magnitude of the coefficient of t in the argument of the sinusoid is proportional to the frequency, so the rank of the waves in order of increasing wavelength is a = b = d < c.

41. From the graph for D_2, $T_2 = \frac{2}{3}$ s and $A = 1.0$ cm. Sine $\lambda_2 = 1.0$ m,

$$D_2(x,t) = (1.0 \text{ cm})\sin(2\pi x - 3\pi t + \phi)$$

From the graph for D_2 we notice that $D_2(1,0) = 1.0$ cm. Therefore,

$$D_2(1,0) = (1.0 \text{ cm})\sin(2\pi + \phi) = 1.0 \text{ cm}$$

This gives $\phi = \pi/2$ rad, and the equation for D_2 is

$$D_2(x,t) = (1.0 \text{ cm})\sin\left(2\pi x - 3\pi t + \frac{\pi}{2}\right)$$

From the graph of D1, $T_1 = 1.0$ s. Since $v = \frac{\lambda_1}{T_1} = \frac{\lambda_2}{T_2}$, we get

$$\lambda_1 = T_1 \frac{\lambda_2}{T_2} = \frac{3}{2} \text{ m}$$

Therefore,

$$D_1(x,t) = (1.0 \text{ cm})\sin\left(\frac{4\pi}{3}x - 2\pi t + \phi\right)$$

From the graph, $D_1(1,0) \approx 0.3$ cm. Therefore,

$$D_1(1,0) = (1.0 \text{ cm})\sin(4.5 + \phi) = 0.3 \text{ cm}$$

This gives $\phi \approx -3.88$ rad or $\phi \approx -1.35$ rad. The value $\phi \approx -1.35$ rad correctly describes the shape of $D_1(1,t)$. Therefore,

$$D_1(x,t) = (1.0 \text{ cm})\sin\left(\frac{4\pi}{3}x - 2\pi t - 1.35\right)$$

43. (a) $\lambda = \dfrac{2\pi}{k} = \dfrac{2\pi}{10.0} = 0.63$ m

    $f = \dfrac{\omega}{2\pi} = \dfrac{7.50}{2\pi} = 1.2$ Hz

    $v = \dfrac{\omega}{k} = \dfrac{7.50}{10.0} = 0.75$ m/s

    (b) $D(x,t) = (0.05 \text{ m})\sin(10.0x - 7.50t)$

45. (a) $\lambda = \dfrac{2\pi}{k} = \dfrac{2\pi}{2\pi/3.0} = 3$ m

    $f = \dfrac{\omega}{2\pi} = \dfrac{2\pi/6.0}{2\pi} = 0.2$ Hz

    $v = \dfrac{\omega}{k} = \dfrac{2\pi/6.0}{2\pi/3.0} = 0.5$ m/s

    (b) The wave travels to the left.

    (c)

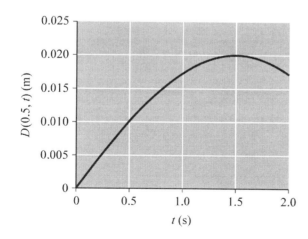

(d) $u(x,t) = \dfrac{\partial}{\partial t} D(x,t)$

$$u(x,t) = (0.02 \text{ m/s})\left[\cos\left(\dfrac{2\pi}{3.0}x + \dfrac{2\pi}{6.0}t - \dfrac{\pi}{3}\right)\right] \times \dfrac{2\pi}{6.0}$$

$$u(0.5 \text{ m}, 2.0 \text{ s}) = (0.02 \text{ m/s})\left[\cos\left(\dfrac{2\pi}{3.0}(0.5) + \dfrac{2\pi}{6.0}(2.0) - \dfrac{\pi}{3}\right)\right] \times \dfrac{2\pi}{6.0}$$

$$= (0.02 \text{ m/s})\left[\cos\left(\dfrac{2\pi}{3}\right)\right] \times \dfrac{2\pi}{6.0} = -0.01 \text{ m/s}$$

47. (a) Reading from the graph, the wavelength of the wave is 8 m.

(b) $v = f\lambda = (5.0 \text{ Hz})(8 \text{ m}) = 40 \text{ m/s}$

(c) If the wave is modelled by a cosine function, its phase constant is zero; if is modelled by a sine function, its phase constant is $\pi/2$.

(d) $D(x,t) = 1.5\sin\left(\dfrac{\pi}{4}x - 10\pi t + \dfrac{\pi}{2}\right)$

49. (a) Reading from the graph, the period is $T = 2$ s, so $f = \dfrac{1}{T} = \dfrac{1}{2} = 0.5 \text{ Hz}$.

(b) $v = f\lambda$, so $\lambda = \dfrac{v}{f} = \left(\dfrac{4.0 \text{ m/s}}{0.5 \text{ s}^{-1}}\right) = 8 \text{ m}$

(c) The equation for the wave can thus be written as

$$D(x,t) = (1.5 \text{ m})\sin\left(\dfrac{\pi}{4}x - \pi t + \phi\right)$$

To determine the phase constant ϕ, notice from the graph that $D(2.5, 0) = -0.5 \text{ m}$. Therefore,

$\sin(1.96 + \phi) = -0.3$

The two possible values of ϕ are given by

$\sin(1.96 + \phi) = -0.3$
$\quad 1.96 + \phi = \sin^{-1}(-0.3)$
$\quad\quad \phi = \sin^{-1}(-0.3) - 1.96 = -2.3 \text{ rad}$

and

$\sin(\pi - 1.96 - \phi) = -0.3$
$\quad \pi - 1.96 - \phi = \sin^{-1}(-0.3)$
$\quad\quad \phi = 1.5 \text{ rad}$

The given shape for the $D(2.5, t)$ plot corresponds to $\phi = 1.5$ rad.

(d) $D(x,t) = (1.5 \text{ m}) \sin\left(\dfrac{\pi}{4}x - \pi t + 1.5\right)$

51. (a) From the first graph, $\lambda = 8 \text{ m}$, and from the second graph, $T = 1 \text{ s}$. Thus, the equation for the wave is given by (since the direction of wave motion is not given, we keep both signs in the second term of the argument of the sine function)

$$D(x,t) = (1.5 \text{ m}) \sin\left(\dfrac{\pi}{4}x \mp 2\pi t + \phi\right)$$

From the first graph, note that $D(0,1) = 1.5 \text{ m}$. This gives, for both signs of the $2\pi t$ term, $\phi = \dfrac{\pi}{2}$ rad. Therefore,

$$D(x,t) = (1.5 \text{ m}) \sin\left(\dfrac{\pi}{4}x \mp 2\pi t + \dfrac{\pi}{2}\right)$$

(b) $D(x,t) = (1.5 \text{ m}) \sin\left(\dfrac{\pi}{4}x \pm 2\pi t + \dfrac{\pi}{2}\right)$

53. (a) From the displacement versus position graph, $\lambda = 3 \text{ m}$ and $A = 0.10 \text{ m}$. From the displacement versus time graph, $T = 6 \text{ s}$. Thus, the equation of the wave is

$$D(x,t) = (0.10 \text{ m}) \sin\left(\dfrac{2\pi}{3}x - \dfrac{2\pi}{6}t + \phi\right) = (0.10 \text{ m}) \sin\left(\dfrac{2\pi}{3}x - \dfrac{\pi}{3}t + \phi\right)$$

From the displacement versus time graph, $D(1, 0) = 0.02 \text{ m}$. Therefore,

$$(0.10 \text{ m}) \sin\left(\dfrac{2\pi}{3} + \phi\right) = 0.02 \text{ m}$$

$$\sin\left(\dfrac{2\pi}{3} + \phi\right) = 0.20$$

$$\dfrac{2\pi}{3} + \phi = 0.201 \text{ or } 2.94$$

$$\phi = -1.9 \text{ or } 0.85$$

The value of the phase constant $\phi = -1.9$ rad correctly gives, as read from the displacement versus time graph, $D(1,2) \approx -0.09 \text{ m}$. Thus,

$$D(x,t) = (0.10 \text{ m}) \sin\left(\dfrac{2\pi}{3}x - \dfrac{\pi}{3}t - 1.9\right)$$

As an additional check, from the above equation, $D(1,1) = -0.07 \text{ m}$, which is consistent with the value given in the displacement versus position graph.

(b) The wave travels to the right at a speed of $v = \dfrac{\omega}{k} = \dfrac{\pi/3}{2\pi/3} = \dfrac{1}{2}$ m/s ≈ 0.50 m/s.

55. (a) The phase constant is 0.2 rad.

(b) The phase of the wave is $4.0x - 0.5t + 0.2 = 4.0(0.5) - 0.5(2.0) + 0.2 = 1.2$ rad.

(c) Yes, the phase changes linearly with time for each fixed value of x.

(d) The phase difference is
$$(4.0x_2 - 0.5t_2 + 0.2) - (4.0x_1 - 0.5t_1 + 0.2) = 4.0(x_2 - x_1) - 0.5(t_2 - t_1) + 0.2 - 0.2$$
$$= 4.0(0.1) - 0.5(0) = 0.4 \text{ rad}$$

(e) Yes; the phase difference is the same for all fixed times for the given difference in x-values.

(f) For a fixed time, the phase difference between two points that are one wavelength apart is
$$4.0(x_2 - x_1) = 4.0 \dfrac{2\pi}{4.0} = 2\pi \text{ rad}$$

which is equivalent to 0 rad.

57. $P = \dfrac{1}{2} \mu v \omega^2 A^2$

$= \dfrac{1}{2} \mu \sqrt{\dfrac{T}{\mu}} (2\pi f)^2 A^2$

$= \sqrt{T\mu} \, 2\pi^2 f^2 A^2$

$= \sqrt{(200.0)(0.1000)} \, 2\pi^2 (10.0)^2 (0.010)^2$

$= 0.88$ W

59. (a)

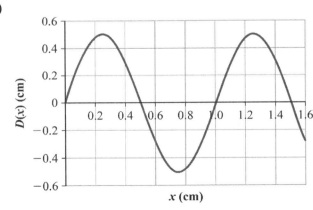

(b) $A = 0.5$ m, $\lambda = 1$ m

(c) The two component waves have the same frequency and wavelength, so they have the same speed. Consequently, the superposition of the two waves also has the same speed.

(d) The resultant wave is a travelling wave, because the two component waves travel in the same direction.

61.

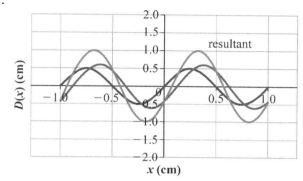

From the graph you can read the phase constant of approximately -0.4 rad.

63. Using Equation 14-40,

$$2A\cos\left(\frac{\phi}{2}\right) = \frac{A}{2}$$

$$\cos\left(\frac{\phi}{2}\right) = \frac{1}{4}$$

$$\frac{\phi}{2} = \cos^{-1}\left(\frac{1}{4}\right)$$

$$\frac{\phi}{2} = 1.3181$$

$$\phi = 2.636 \text{ rad}$$

The phase difference between the resultant wave and either of the two component waves is $\frac{\phi}{2} = 1.318$ rad

65. (a) The minimum amplitude, which occurs when the two waves are out of phase by half a cycle, is $|A_1 - A_2|$.

(b) The maximum amplitude, which occurs when the two waves are in phase, is $A_1 + A_2$.

67. Using the identity

$$\sin(a) + \sin(b) = 2\cos\left(\frac{a-b}{2}\right)\sin\left(\frac{a+b}{2}\right)$$

and letting a and b represent the arguments of the constituent waves, the resultant wave can be expressed as

$$D(x,t) = 2A\cos\left(\frac{[kx-\omega t+\phi_1]-[kx-\omega t+\phi_2]}{2}\right)\sin\left(\frac{[kx-\omega t+\phi_1]+[kx-\omega t+\phi_2]}{2}\right)$$

$$= 2A\cos\left(\frac{\phi_1-\phi_2}{2}\right)\sin\left(\frac{2kx-2\omega t+\phi_1+\phi_2}{2}\right)$$

$$= 2A\cos\left(\frac{\phi_1-\phi_2}{2}\right)\sin\left(kx-\omega t+\frac{\phi_1+\phi_2}{2}\right)$$

69. (a) For a hard reflection, the wave is inverted and travels in the opposite direction; therefore, its wave function is

    $D(x,t) = 0.2\sin(3x-4t+\pi)$

    (b) For a soft reflection, the wave is *not* inverted and travels in the opposite direction; therefore, its wave function is

    $D(x,t) = 0.2\sin(3x-4t)$

71. (a) The frequency of the standing wave is $\frac{20\pi}{2\pi} = 10$ Hz and the wavelength is $\frac{2\pi}{0.6} = 10.5$ m.

    (b) The distance between consecutive nodes is half a wavelength, which is 5.2 m, and the distance between a node and an adjacent antinode is a quarter of a wavelength, which is 2.6 m.

    (c) This standing wave is produced by two travelling waves of the same wavelength (0.6 m) and frequency (10 Hz) that are moving in opposite directions. The amplitude of each travelling wave is 1.50 cm/2 = 0.75 m. The equations for the two waves are given by

    $D_1(x,t) = (0.75\text{ cm})\sin(0.6x-20\pi t)$
    $D_2(x,t) = (0.75\text{ cm})\sin(0.6x+20\pi t)$

    (d) $D(0.20, 3.0) = (1.50\text{ cm})\sin(0.6(0.20))\cos(20\pi(3.0))$
    $= (1.50\text{ cm})\sin(0.12)\cos(60\pi)$
    $= (1.50\text{ cm})(0.1197)(1)$
    $= 0.18$ cm

73. (a) $D(x,t) = (2.0\text{ mm})\sin(\pi x)\cos(0.5\pi t)$

    (b) The first three nodes on either side of $x = 0$ are located at $x = \pm 1$ m, ± 2 m, ± 3 m. The first three antinodes on either side of $x = 0$ are at $x = \pm 0.5$ m, ± 1.5 m, ± 2.5 m.

(c) Based on the result of part (b), the distance between two consecutive nodes is 1.0 m.

75. Using the identity

$$\sin(a) + \sin(b) = 2\cos\left(\frac{a-b}{2}\right)\sin\left(\frac{a+b}{2}\right)$$

and letting a and b represent the arguments of the constituent waves, the resultant wave can be expressed as

$$D(x,t) = 2A\cos\left(\frac{[kx-\omega t + \phi_1]-[kx+\omega t + \phi_2]}{2}\right)\sin\left(\frac{[kx-\omega t + \phi_1]+[kx+\omega t + \phi_2]}{2}\right)$$

$$= 2A\cos\left(-\omega t + \frac{\phi_1 - \phi_2}{2}\right)\sin\left(\frac{2kx + \phi_1 + \phi_2}{2}\right) = 2A\cos\left(\omega t - \frac{\phi_1 - \phi_2}{2}\right)\sin\left(kx + \frac{\phi_1 + \phi_2}{2}\right)$$

77. $f = \frac{m}{2L}\sqrt{\frac{T}{\mu}} = \frac{1}{2(0.500)}\sqrt{\frac{T}{\mu}} = \sqrt{\frac{T}{\mu}} = 260.0$ Hz

When the tension is increased by 4%, the new frequency is

$$f = \sqrt{\frac{1.04T}{\mu}} = \sqrt{1.04}\sqrt{\frac{T}{\mu}} = 1.0198(260.0) = 265 \text{ Hz}$$

79. (a) The longest-wavelength standing wave possible is 4.0 m, because at least half a wavelength must fit into the length of the string.

(b) $f = \frac{v}{2L} = \frac{200.0}{2(2.0)} = 50.$ Hz

(c) No; the allowed frequencies are whole-number multiples of the fundamental frequency.

(d) The frequency of the second harmonic is $2(50. \text{ Hz}) = 1.0 \times 10^2$ Hz, and the frequency of the fourth harmonic is $4(50. \text{ Hz}) = 2.0 \times 10^2$ Hz.

second harmonic   fourth harmonic

81. (a) The fundamental frequency is

$$f = \frac{1}{2L}\sqrt{\frac{T}{\mu}} = \frac{1}{2(1.000 \text{ m})}\sqrt{\frac{25.00 \text{ N}}{0.000\ 650 \text{ kg/m}}} = 98.1 \text{ Hz}$$

(b) The frequency of the second harmonic is $2(98.1 \text{ Hz}) = 196$ Hz, and the frequency of the third harmonic is $3(98.1 \text{ Hz}) = 294$ Hz.

(c) The wave speed is

$$v = \sqrt{\frac{T}{\mu}} = \sqrt{\frac{25.00 \text{ N}}{0.000\,650 \text{ kg/m}}} = 196 \text{ m/s}$$

(d) The new fundamental frequency is

$$f = \frac{1}{2L}\sqrt{\frac{T}{\mu}} = \frac{1}{2(1.000 \text{ m})}\sqrt{\frac{35.00 \text{ N}}{0.000\,650 \text{ kg/m}}} = 116 \text{ Hz}$$

83. We need to divide the interval between an octave into 10 semitones. If f is the starting frequency, then the frequency an octave higher is $2f$. Since

$$(2^{1/10})^{10} = 2$$

the frequency difference between adjacent semitones is $2^{1/10} \approx 1.071\,77$. Frequencies of semitones between 440 Hz and 880 Hz are as follows:

Semitone	Frequency
1st	$2^{1/10} \times 440$ Hz $= 470$ Hz
2nd	$2^{2/10} \times 440$ Hz $= 510$ Hz
3rd	$2^{3/10} \times 440$ Hz $= 540$ Hz
4th	$2^{4/10} \times 440$ Hz $= 580$ Hz
5th	$2^{5/10} \times 440$ Hz $= 620$ Hz
6th	$2^{6/10} \times 440$ Hz $= 670$ Hz
7th	$2^{7/10} \times 440$ Hz $= 710$ Hz
8th	$2^{8/10} \times 440$ Hz $= 770$ Hz
9th	$2^{9/10} \times 440$ Hz $= 820$ Hz

85. Given quantities are

$f_1 = 220$ Hz

$L = 69.0$ cm $= 0.690$ m

$m = 1.90$ g $= 1.90 \times 10^{-3}$ kg

(a) Since

$$f_1 = \frac{1}{2L}\sqrt{\frac{T_s}{\mu}}$$

solving for T_s gives

$$T_s = 4\mu L^2 f_1^2 = 4\left(\frac{1.90\times 10^{-3}\text{ kg}}{0.69\text{ m}}\right)(0.690\text{ m})^2(220\text{ Hz})^2 = 254\text{ N}$$

(b) A frequency of two semitones above 220 Hz corresponds to 880 Hz. Let the length of the vibrating section of the string corresponding to this frequency be L'. Since the wave speed does not change, we must have

$$f_1 \times L = f_1' \times L'$$

where $f_1' = 880.0$ Hz. Therefore,

$$L' = \left(\frac{f_1}{f_1'}\right)L = \left(\frac{220\text{ Hz}}{880\text{ Hz}}\right)L = \frac{1}{4}L = 17.25\text{ cm}$$

(c) $f_3 = 3f_1 = 3\times 220\text{ Hz} = 660\text{ Hz}$

87. In this case we are given the following quantities:

Diameter of the wire $= d = 0.054$ inches $= 0.054$ inches $\times \dfrac{2.54\text{ cm}}{\text{inch}} \times \dfrac{1\text{ m}}{100\text{ cm}} = 1.372\times 10^{-3}$ m

Tension in the wire $= T_s = 220$ N

Length of the wire $= L = 65$ cm $= 0.65$ m

Density of the wire $= \rho = 8.5$ g/cm³ $= 8.5\times 10^3$ kg/m³

(a) To determine the frequency of the fundamental harmonic for this wire, we need to calculate its linear mass density, μ:

$$\mu = \pi\left(\frac{d}{2}\right)^2 \rho = \pi\left(\frac{1.372\times 10^{-3}\text{ m}}{2}\right)^2(8.5\times 10^3\text{ kg/m}^3) = 1.257\times 10^{-2}\text{ kg/m}$$

The frequency of the fundamental harmonic is

$$f_1 = \frac{1}{2L}\sqrt{\frac{T_s}{\mu}} = 102\text{ Hz}$$

(b) Let $f' = 2f_1 = 204$ Hz. The wire is still vibrating in the first harmonic with this new frequency. As the tension and the linear mass density do not change, the new length, L', of the wire is given by

$$f' = \frac{1}{2L'}\sqrt{\frac{T_s}{\mu}} = \left(\frac{L}{L'}\right)f$$

Therefore,

$$L' = L\left(\frac{f}{f'}\right) = \frac{L}{2} = 32.5 \text{ cm}$$

89. (a) $f_a = \dfrac{1}{2 \times 0.5}\sqrt{\dfrac{100}{\mu}} = \dfrac{10}{\sqrt{\mu}}$

(b) $f_b = \dfrac{2}{2 \times 0.50}\sqrt{\dfrac{200}{\mu}} = \dfrac{20\sqrt{2}}{\sqrt{\mu}}$

(c) $f_c = \dfrac{1}{1}\sqrt{\dfrac{100}{4\mu}} = \dfrac{5}{\sqrt{\mu}}$

(d) $f_d = \dfrac{3}{2 \times 0.75}\sqrt{\dfrac{200}{\mu}} = \dfrac{20\sqrt{2}}{\sqrt{\mu}}$

Therefore, $f_c < f_a < f_b = f_d$.

Chapter —INTERFERENCE AND SOUND

In many of the solutions for this chapter, units are omitted in calculations for simplicity.

1. True. Sound waves are longitudinal.

3. The speed of a wave is the product of its frequency and its wavelength, and the speed depends on properties of the medium. Thus, if the frequency doubles, then the wavelength is divided by two or, equivalently, is multiplied by a factor of 0.5.

5. True. As with light waves, angles of incidence and reflection are equal for sound waves.

7. True. Confining a wave between boundaries results in interference that produces a standing wave, provided that the wave has an appropriate wavelength.

9. False. It is possible to have a number of different standing waves exist at the same time in an air column; they will combine to produce one sound.

11. Consider two sources of intensities I_1 and I_2, respectively. The corresponding sound intensities are β_1 and β_2. Then

$$\beta_1 = 10\log_{10}\left(\frac{I_1}{I_0}\right) \quad \text{and} \quad \beta_2 = 10\log_{10}\left(\frac{I_2}{I_0}\right)$$

$$\beta_2 - \beta_1 = 10\log_{10}\left(\frac{I_2}{I_0}\right) - 10\log_{10}\left(\frac{I_1}{I_0}\right) = 10\left[\log_{10}\left(\frac{I_2}{I_0}\right) - \log_{10}\left(\frac{I_1}{I_0}\right)\right]$$

$$= 10\left[\log_{10}\left(\frac{I_2}{I_0} \div \frac{I_1}{I_0}\right)\right] = 10\left[\log_{10}\left(\frac{I_2}{I_0} \times \frac{I_0}{I_1}\right)\right] = 10\left[\log_{10}\left(\frac{I_2}{I_1}\right)\right]$$

Now, if $I_2 = 2I_1$, then

$$\beta_2 - \beta_1 = 10\log_{10}\left(\frac{2I_1}{I_1}\right) = 10\log_{10}(2) = 10(0.301) = 3.01$$

Thus, the increase in sound intensity level is 3.01 dB (to three significant digits) when the sound intensity is doubled.

13. c. Sound waves can interfere both spatially and temporally.

15. $v = f\lambda$

$v = 343$ m/s for air at 20 °C

$$\lambda = \frac{v}{f} = \frac{343 \text{ m/s}}{1000 \text{ s}^{-1}} = 0.343 \text{ m} = 0.3 \text{ m to one significant digit}$$

17. The speed of sound in water at 20 °C is 1482 m/s. Thus,

$$\Delta y = v\Delta t \rightarrow \Delta t = \frac{\Delta y}{v}$$

$$\Delta t = \frac{30 \text{ m} + 30 \text{ m}}{1482 \text{ m/s}} = 0.04 \text{ s}$$

19. $s(x,t) = s_m \cos(kx - \omega t) = s_m \cos\left(\frac{2\pi}{\lambda}x - 2\pi f t\right)$

Now, $f = 60$ Hz, so $2\pi f = 377$ s^{-1}, or 400 s^{-1} to one significant digit. For sound in air, $v = 343$ m/s, so

$$k = \frac{2\pi}{\lambda} = \frac{2\pi f}{v} = \frac{2\pi(60 \text{ s}^{-1})}{343 \text{ m/s}} = 1.1 \text{ m}^{-1} = 1 \text{ m}^{-1} \text{ to one significant digit}$$

Thus, the displacement amplitude can be expressed as

$s(x,t) = s_m \cos(x - 400t)$

21. The amplitude of pressure variations is $\Delta p_m = Bks_m$, where s_m is the amplitude of displacement variations, and the bulk modulus of air is $B = 1.01 \times 10^5$ Pa. Now

$$k = \frac{2\pi}{\lambda} = \frac{2\pi f}{v} = \frac{2\pi(1000 \text{ Hz})}{343 \text{ m/s}} = 18.32 \text{ m}^{-1} = 20 \text{ m}^{-1} \text{ to one significant digit}$$

$$s_m = \frac{\Delta p_m}{Bk} = \frac{20 \text{ N/m}^2}{(1.01 \times 10^5 \text{ Pa})(18.32 \text{ m}^{-1})} = 1.1 \times 10^{-5} \text{ m} = 1 \times 10^{-5} \text{ m to one significant figure}$$

23. $\Delta p_m = Bks_m$, so when the displacement amplitude doubles, so does the pressure amplitude.

25.

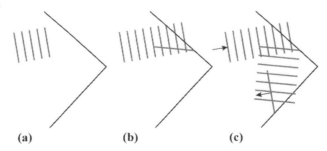

(a) (b) (c)

27. $f_1 = \frac{v}{2L} = \frac{343 \text{ m/s}}{2(0.75 \text{ m})} = 230$ Hz

29. The maxima in the microphone occur at the frequencies where the tube supports standing waves. For a tube that is open at both ends, the frequency of the nth harmonic is given by

$f_n = \frac{nv}{2L}$. We might be tempted to assume that the lowest frequency in the data set is the fundamental, corresponding to $n = 1$. If we did that, we would find

$$L = \frac{v}{2f_1} = \frac{343 \text{ m}\cdot\text{s}^{-1}}{2(110 \text{ s}^{-1})} = 1.56 \text{ m}$$

However, a better way of looking at the data is to examine the change in frequency between observations. We can readily see that the frequency increases by 80 Hz between maxima. The change in frequency given by the formula is

$$\Delta f = f_{n+1} - f_n = \frac{(n+1)v}{2L} - \frac{nv}{2L} = \frac{v}{2L}$$

or

$$L = \frac{v}{2\Delta f} = \frac{343 \text{ m}\cdot\text{s}^{-1}}{2(80 \text{ s}^{-1})} = 2.14 \text{ m}$$

From this we conclude that if the measurements had gone below 100 Hz, a maximum would have been observed at 80 Hz as well, and that would be the fundamental.

31. The beat frequency is 296 Hz − 294 Hz = 2 Hz.

33. The beat frequency is 440 Hz − 350 Hz = 90 Hz.

35. $\beta_1 = 10 \log\left(\frac{I_1}{I_0}\right) = 95$ dB

$\beta_2 = 10 \log\left(\frac{2I_1}{I_0}\right) = 10\left[\log(2) + \log\left(\frac{I_1}{I_0}\right)\right] = 10 \log(2) + 10 \log\left(\frac{I_1}{I_0}\right)$
$= 10 \log(2) + 95 \text{ dB} = 3.01 \text{ dB} + 95 \text{ dB} = 98 \text{ dB}$

37. We first determine the intensity of the lion's roar at 40 m. Recall that

$$\beta(40) = 10 \log\left(\frac{I(40)}{I_0}\right)$$

$$60 = 10 \log\left(\frac{I(40)}{I_0}\right)$$

$$10^6 = \frac{I(40)}{I_0}$$

$$I(40) = 10^6 I_0$$

Now, we know that the intensity, $I(r)$, a distance r from an isotropic sound source of power P is given by

$$I(r) = \frac{P}{4\pi r^2}$$

We know what $I(40)$ is, so we could use the equation above to find the total power, P, radiated by the lion and then use that to find $I(10)$. However, we can skip a step by noting that the ratio of the intensities at two different distances is simply the inverse ratio of the distances squared as follows:

$$\frac{I(r_2)}{I(r_1)} = \frac{\frac{P}{4\pi r_2^2}}{\frac{P}{4\pi r_1^2}} = \left(\frac{r_1}{r_2}\right)^2$$

$$I(r_2) = I(r_1)\left(\frac{r_1}{r_2}\right)^2$$

Thus,

$$I(10) = I(40)\left(\frac{40}{10}\right)^2 = 10^6 \times I_0 \times 16$$

$$\beta(10) = 10\log\left(\frac{I(10)}{I_0}\right) = 10\log\left(\frac{16 \times 10^6 \times I_0}{I_0}\right) = 10\log(16 \times 10^6) = 72 \text{ dB}$$

39. $f_r = \frac{v \pm v_r}{v \mp v_s}f_s = \frac{v}{v + v_s}f_s = \frac{1482}{1482 + 0.5}(10.0 \text{ MHz}) = 9.9966 \text{ MHz}$

The frequency shift is 10.0 MHz − 9.9966 MHz = 3.4 kHz. Note that we needed to keep more significant digits than were provided in the question to see this difference.

41. We need to be very careful in this problem to clearly identify the source, the receiver, and how each is moving. In all cases the bat first emits the sound. However, the sound is then reflected off the moth. So, we will first determine the frequency received by the moth with the bat as the source. We will then use that frequency as the source frequency to determine the frequency perceived by the bat as the receiver. The equation we will be using is Equation 15-30

$$f_r = \frac{v \pm v_r}{v \mp v_s}f_s$$

where the sign convention is upper when the motion is toward and lower when the motion is away.

(a) In this case, the bat is flying toward the stationary moth. Because the bat is approaching the moth, the moth will detect a frequency that is Doppler shifted up due to the speed of the approaching bat. In the first instance, we have a moving source and a stationary receiver, so we select the upper sign for the bat speed and use

$$f_{r,moth} = \frac{v}{v - v_{bat}} f_{t,bat} = \frac{343 \text{ m·s}^{-1}}{343 \text{ m·s}^{-1} - 7.00 \text{ m·s}^{-1}} 82.0 \text{ kHz} = 83.7 \text{ kHz}$$

This frequency now becomes the frequency emitted by the moth back to the bat:

$$f_{r,bat} = \frac{v + v_{bat}}{v} f_{t,moth} = \frac{343 \text{ m·s}^{-1} + 7.00 \text{ m·s}^{-1}}{343 \text{ m·s}^{-1}} 83.7 \text{ kHz} = 85.4 \text{ kHz}$$

(b) Both the bat and the moth are now moving. The bat is flying faster than the moth, but they are both flying in the same direction, so we would anticipate a slightly smaller Doppler shift this time. We will use the same strategy as last time. Here we need to be very careful with our sign choices. The bat is flying toward the moth, so we use the upper sign for the bat speed. However, the moth is flying away from the bat, so we use the lower sign for the moth speed:

$$f_{r,moth} = \frac{v - v_{moth}}{v - v_{bat}} f_{t,bat} = \frac{343 \text{ m·s}^{-1} - 1.00 \text{ m·s}^{-1}}{343 \text{ m·s}^{-1} - 7.00 \text{ m·s}^{-1}} 82.0 \text{ kHz} = 83.5 \text{ kHz}$$

We now use this frequency as the source frequency of the moth to determine the frequency received by the bat:

$$f_{r,bat} = \frac{v + v_{bat}}{v + v_{moth}} f_{t,moth} = \frac{343 \text{ m·s}^{-1} + 7.00 \text{ m·s}^{-1}}{343 \text{ m·s}^{-1} + 1.00 \text{ m·s}^{-1}} 83.5 \text{ kHz} = 85.0 \text{ kHz}$$

(c) In this case, the moth is (foolishly) flying toward the bat at 2.00 m/s and the bat is flying toward the moth, so we use the upper sign in both cases. We will again employ the two-step strategy:

$$f_{r,moth} = \frac{v + v_{moth}}{v - v_{bat}} f_{s,bat} = \frac{343 \text{ m·s}^{-1} + 2.00 \text{ m·s}^{-1}}{343 \text{ m·s}^{-1} - 7.00 \text{ m·s}^{-1}} 82.0 \text{ kHz} = 84.2 \text{ kHz}$$

We now use this frequency as the source frequency from the moth to find the frequency received by the bat:

$$f_{r,bat} = \frac{v + v_{bat}}{v + v_{moth}} f_{t,moth} = \frac{343 \text{ m·s}^{-1} + 7.00 \text{ m·s}^{-1}}{343 \text{ m·s}^{-1} - 2.00 \text{ m·s}^{-1}} 84.2 \text{ kHz} = 86.4 \text{ kHz}$$

43. Light takes only very slightly more than 0 s to reach you, so this time is negligible. The distance travelled by sound in 4.0 s is (343 m/s)(4.0 s) = 1.4 km.

45. $s_1(x,t) = 10\cos(kx - \omega t)$ and $s_2(x,t) = 20\cos\left(kx - \omega t \pm \dfrac{\pi}{4}\right)$

Thus,

$$s_1(x,t) + s_2(x,t) = 10\cos(kx - \omega t) + 20\cos\left(kx - \omega t \pm \dfrac{\pi}{4}\right)$$

$$= 10\cos(kx - \omega t) + 20\cos(kx - \omega t)\cos\left(\dfrac{\pi}{4}\right) \mp 20\sin(kx - \omega t)\sin\left(\dfrac{\pi}{4}\right)$$

$$= (10 + 10\sqrt{2})\cos(kx - \omega t) \mp 10\sqrt{2}\sin(kx - \omega t) \quad \text{(F)}$$

$$= A\cos(kx - \omega t - \theta), \text{ for some constants } A \text{ and } \theta$$

Expanding the expression on the right side of the previous line, we obtain

$A\cos(kx - \omega t)\cos\theta + A\sin(kx - \omega t)\sin\theta$

Comparing with (F), we obtain

$A\cos\theta = 10 + 10\sqrt{2}$ and $A\sin\theta = \mp 10\sqrt{2}$

Squaring and adding the two equations in the previous line, we obtain

$A^2\cos^2\theta + A^2\sin^2\theta = [10(1 + \sqrt{2})]^2 + (10\sqrt{2})^2$

$A^2(\cos^2\theta + \sin^2\theta) = 100(1 + 2\sqrt{2} + 2 + 2)$

$A^2 = 100(5 + 2\sqrt{2})$

$A = 10\sqrt{5 + 2\sqrt{2}} = 27.979 \text{ nm} = 28 \text{ nm} = 30 \text{ nm}$

47. $I = I_0 10^{\beta/10} = 10^{-12} \text{ W/m}^2 \times 10^{65/10} = 10^{-5.5} \text{ W/m}^2 = 3.16 \times 10^{-6} \text{ W/m}^2$

$= 3.2 \times 10^{-6} \text{ W/m}^2$

$P = (4\pi r^2)I = 4\pi(23 \text{ m})^2(3.16 \times 10^{-6} \text{ W/m}^2) = 0.021 \text{ W} = 21 \text{ mW}$

49. The speed of the car is 130 km/h = 36.1 m/s. We have

$$f_r = \dfrac{v \pm v_r}{v \mp v_s} f_s$$

Thus, the frequency received by the moving car is

$$f_r = \dfrac{c - 36.1 \text{ m/s}}{c}(10 \text{ GHz})$$

This is the same frequency as that of the reflected waves, which are received back at the radar unit with a frequency of

$$f_r = \left(\frac{c}{c+36.1 \text{ m/s}}\right)\left(\frac{c-36.1 \text{ m/s}}{c}\right)(10 \text{ GHz}) = \left(\frac{c-36.1 \text{ m/s}}{c+36.1 \text{ m/s}}\right)(10 \text{ GHz})$$

$$= \left(\frac{3.00\times 10^8 \text{ m/s} - 36.1 \text{ m/s}}{3.00\times 10^8 \text{ m/s} + 36.1 \text{ m/s}}\right)(10 \text{ GHz}) = 9.999\,997\,593 \text{ GHz}$$

The beat frequency is $f_s - f_r = 2400$ Hz $= 2.4$ kHz.

51. $G_4 \to 392.00$ Hz
    $C_5 \to 523.25$ Hz
    $E_5 \to 659.26$ Hz
    $G_5 \to 783.99$ Hz

    $392.00 = kf_1$
    $523.25 = (k+1)f_1$
    $f_1 = 523.25 - 392.00 = 131.25$ Hz

    Similarly,

    659.26 Hz $- 523.25$ Hz $= 136.01$ Hz $= f_1$
    783.99 Hz $- 659.26$ Hz $= 124.73$ Hz $= f_1$

    The average value is

    $$f_1 = \frac{124.73 \text{ Hz} + 136.01 \text{ Hz} + 131.25 \text{ Hz}}{3} = 131 \text{ Hz}$$

    Thus,

    $$f_1 = \frac{v}{2L}, \text{ so } L = \frac{v}{2f_1} = \frac{343 \text{ m/s}}{2(131 \text{ Hz})} = 1.3 \text{ m}$$

53. $f_1 = \dfrac{v}{4L}$

    $v = 4Lf_1 = \dfrac{4L}{T_1} = \dfrac{4(270 \text{ km})}{12.5 \text{ h}} = 86.4$ km/h $= 24$ m/s

55. $I = I_0 10^{\beta/10} \Rightarrow \dfrac{I_2}{I_1} = 10^{(\beta_2 - \beta_1)/10}$

    A reduction in sound level by 60 dB corresponds to a reduction in intensity by a factor of $10^{-60/10} = 10^{-6}$. Thus, the new sound intensity is $0.2 \ \mu W/m^2$.

57. Let y represent the depth of the well. The time needed for the stone to drop is t, where

$$y = \frac{1}{2}gt_1^2 \Rightarrow t_1 = \sqrt{\frac{2y}{g}}$$

The additional time needed for the sound of the splash to reach the top of the well is t_2, where

$$y = vt_2$$

$$t_2 = \frac{y}{343}$$

The total time is 2.8 s, so $t_1 + t_2 = 2.8$ s. Thus,

$$\sqrt{\frac{2y}{g}} + \frac{y}{343 \text{ m/s}} = 2.8 \text{ s}$$

$$\sqrt{\frac{2y}{g}} = 2.8 \text{ s} - \frac{y}{343 \text{ m/s}}$$

$$\frac{2y}{g} = \left(2.8 \text{ s} - \frac{y}{343 \text{ m/s}}\right)^2 = (2.8 \text{ s})^2 - \frac{5.6 \text{ s}}{343 \text{ m/s}}y + \frac{y^2}{(343 \text{ m/s})^2}$$

$$0 = 2.8^2 - \left(\frac{5.6}{343} + \frac{2}{g}\right)y + \frac{y^2}{343^2}$$

$$y = \frac{\left(\frac{5.6 \text{ s}}{343 \text{ m/s}} + \frac{2}{g}\right) \pm \sqrt{\left(\frac{5.6 \text{ s}}{343 \text{ m/s}} + \frac{2}{g}\right)^2 - \frac{4(2.8 \text{ s})^2}{(343 \text{ m/s})^2}}}{(2/343 \text{ m/s}^2)} = 25\,900 \text{ m or } 35.7 \text{ m}$$

25 900 m is rejected because t_2 by itself would be much greater than 2.8 s. Thus, the depth of the well is 36 m.

59. Because the sound is loudest when you are equidistant from the sources, the sources are in phase. The condition for destructive interference is

$$\Delta d = \left(n + \frac{1}{2}\right)\lambda$$

where n is a natural number and λ is the wavelength of waves emitted by the sources. Thus,

$$\lambda = \frac{\Delta d}{n + 1/2}$$

$$\frac{v}{f} = \frac{\Delta d}{n + 1/2}$$

$$f = \frac{v(n+1/2)}{\Delta d} = \frac{343 \text{ m/s}(n+1/2)}{2.8 \text{ m}} \quad f = 122.5(n+1/2) \text{ s}^{-1}$$

If this is the lowest-order minimum, then $n = 0$, and

$$f = \frac{122.5 \text{ Hz}}{2} = 61 \text{ Hz}$$

61. (a) For the fundamental frequency, half a wavelength fits into the length of the rod, so $\lambda_1 = 240$ cm.

    (b) $v = f_1 \lambda_1$

    $$f_1 = \frac{v}{\lambda_1} = \frac{6420 \text{ m/s}}{2.4 \text{ m}} = 2675 \text{ Hz} = 2.7 \text{ kHz}$$

63. Our two waves then become:

    $D_1(0,t) = A\cos(-\omega_1 t + \phi_1)$ and $D_2(0,t) = A\cos(-\omega_2 t + \phi_2)$

    To find the resulting wave, we apply the principle of superposition and add the two waves:

    $$\begin{aligned}D_{Total}(0,t) &= D_1(0,t) + D_2(0,t) \\ &= A\cos(-\omega_1 t + \phi_1) + A\cos(-\omega_2 t + \phi_2) \\ &= A[\cos(-\omega_1 t + \phi_1) + \cos(-\omega_2 t + \phi_2)] \\ &= 2A\cos\left[\frac{1}{2}(-\omega_1 t + \phi_1 - \omega_2 t + \phi_2)\right] \\ &\quad \times \cos\left[\frac{1}{2}(-\omega_1 t + \phi_1 - \omega_2 t + \phi_2)\right] \\ &= 2s_m \cos\left[-\left(\frac{\omega_1 + \omega_2}{2}\right)t + \frac{\phi_1 + \phi_2}{2}\right] \\ &\quad \times \cos\left[-\left(\frac{\omega_1 - \omega_2}{2}\right)t + \frac{\phi_1 - \phi_2}{2}\right]\end{aligned}$$

    To simplify this expression, we define the following quantities:

    Mean angular frequency: $\bar{\omega} = \dfrac{\omega_1 + \omega_2}{2}$

    Angular frequency difference: $\Delta\omega = \dfrac{\omega_1 - \omega_2}{2}$

    Average phase constant: $\bar{\phi} = \dfrac{\phi_1 + \phi_2}{2}$

    Phase constant difference: $\Delta\phi = \dfrac{\phi_1 - \phi_2}{2}$

    We then get

    $$D_{Total}(t) = 2A\cos(-\Delta\omega t + \Delta\phi)\cos(-\bar{\omega}t + \bar{\phi})$$

This is a wave that has a frequency of $\bar{\omega} = \dfrac{\omega_1 + \omega_2}{2}$ that is amplitude modulated at a frequency of $\Delta\omega = \dfrac{\omega_1 - \omega_2}{2}$. Note that this result differs from Equation 15-25 only in that some small phase differences have been introduced.

65. To find the speed, we use

$$u = c\left(\frac{\lambda' - \lambda}{\lambda'}\right) = c\left(\frac{6800.0 \times 10^{-10}\ \text{m}\cdot\text{s}^{-1} - 6562.8 \times 10^{-10}\ \text{m}\cdot\text{s}^{-1}}{6800.0 \times 10^{-10}\ \text{m}\cdot\text{s}^{-1}}\right)$$

$$= 1.05 \times 10^7\ \text{m}\cdot\text{s}^{-1} = 1.05 \times 10^4\ \text{km}\cdot\text{s}^{-1}$$

To find the distance, we use Hubble's law, $u = HD$, rearranged for D to get

$$D = \frac{u}{H} = \frac{1.05 \times 10^4\ \text{km}\cdot\text{s}^{-1}}{70.2\ \text{km}\cdot\text{s}^{-1}\cdot\text{Mpc}^{-1}} = 150.\ \text{Mpc}$$

67. We would expect a long line source to radiate out sound in all directions in planes that are perpendicular to the line source. If we were to walk around the line source, we would hear a constant sound level. Similarly, if we walked parallel to the line source, we would hear a constant sound level. However, if we walked away from the source, the sound level would diminish.

For the case of a point source of sound, we considered a spherical surface of radius r, with the source at the centre of the sphere. All of the power, P, radiated by the point source would have to pass though the surface of the sphere, and everywhere on the surface of the sphere we would measure a uniform intensity, I. The relationship between the total power, measured in watts, and the intensity, measured in watts per square metre, is

$$I = \frac{P}{4\pi r^2}$$

For the line source, we could consider a cylinder, radius r and length ℓ, that is coaxial with and centred on the line. Since the sound waves are propagating radially from the line, they only pass through the walls of the cylinder and not the two ends. Thus, the power radiated by the length of the source inside the cylinder radiates through an area $A = 2\pi r \ell$.

We need to be a little careful when we consider the power in this case. For the point source, we enclosed the entire source with the sphere and thus the total power radiated all came though the surface of the sphere. For the case of the line source, we are only enclosing part of the source and thus only part of the total power that the source radiates will pass through the walls of our cylinder. To sort this out, we will define a linear power density, λ, as the ratio of the total power radiated by the source, P, to the total length of the source, L:

$$\lambda = \frac{P_{\text{total}}}{L}$$

The linear power density will have units of watts per metre. Our cylinder encloses ℓ metres of the source, and thus the power radiated by that segment of the source is

$$P_{segment} = \lambda \ell$$

Thus, the intensity measured on the surface of the cylinder is

$$I = \frac{P_{segment}}{A} = \frac{\lambda l}{2\pi r l} = \frac{\lambda}{2\pi r}$$

From this result, we can see that the intensity decreases as $1/r$ as we move away from a line source of sound. This is one of the reasons the thunder associated with a lightning strike is so loud even when the strike is at a considerable distance.

These ideas are also incorporated into modern loudspeaker design, with a linear array of smaller speakers becoming much more popular.

69. The speed of sound in helium is much greater than the speed of sound in air. Although your mouth and throat are not exactly a tube, the same principles apply to your mouth and throat as do to a tube, albeit in a more complicated way. Thus, the resonant frequencies of your throat and mouth are also proportional to

$$\frac{v}{L}$$

Because the speed of sound is higher in helium, the resonant frequencies are higher for sound travelling in your throat/mouth system when it is filled with helium instead of air.

Chapter —PHYSICAL OPTICS

1. d

3. c. The constructive interference condition for the Michelson interferometer is
$$m\lambda = 2d \quad (1)$$

   In (1), λ denotes the wavelength of the light used in the interference experiment, d is the distance the mirror was moved, and m is the number of constructive interference maxima or bright fringes. If monochromatic light with two wavelengths, $\lambda_1 = 645$ nm and $\lambda_2 = 430$ nm, is used separately, but in the same interference condition (the same value of d), then (1) leads to

   $$m_1 \lambda_1 = m_2 \lambda_2 \Rightarrow m_2 = m_1 \frac{\lambda_1}{\lambda_2} = 64 \times \frac{645\,\text{nm}}{430\,\text{nm}} = 96$$

5. b. Using the constructive interference condition for the double-slit experiment and the small-angle approximation, we can determine that the angle separation between the central spot and the first-order bright fringe on each side is approximately given by (see also (3) and (4) from the solution of question 4)

   $$\theta \approx \frac{\lambda}{d}$$

   Hence, for two different separations between the two slits, d_1 and d_2, and the same wavelength, λ, we can write that

   $$\frac{\theta_1}{\theta_2} = \frac{d_2}{d_1} \Rightarrow d_2 = d_1 \frac{\theta_1}{\theta_2} = s\frac{1.2°}{2.4°} = \frac{s}{2}$$

7. d. The constructive interference condition for a diffraction grating is
   $$d\sin\theta = m\lambda$$

   Therefore, since d is constant, the bright fringes nearest to the central white spot correspond to the smallest wavelength, λ. For the stellar visible light spectrum, the minimum and maximum wavelength values correspond to the colours violet and red, respectively. Since for $m = 0$ the wavelength does not matter, the central spot is white.

9. b. The destructive interference condition for single-slit diffraction is
   $$w\sin\theta = m\lambda$$

   Therefore, for a constant slit width w, a larger wavelength λ will yield a larger angle θ satisfying the equation above, hence a more spread out diffraction pattern in both directions.

11. b. The first order of constructive interference for a light beam of wavelength λ in vacuum normally incident on the soap film with index of refraction n is

    $$2t = \frac{\lambda}{2n} \Rightarrow \lambda = 4nt$$

Hence, as the soap film evaporates, its thickness, t, decreases, and so does the wavelength, λ, of the light for which the interference condition holds.

13. b. The resolution improves if an electromagnetic wave with smaller wavelength λ is used.

15. d. The Fraunhofer condition requires that the distance to the observing screen be very large compared to the dimensions of the object producing the interference pattern, the double slit in this case.

17. a. Since this is going to a medium with a lower index of refraction, it is a soft reflection and there is no phase change for the reflected wave. The transmitted wave never has a phase change.

19. b. The reflection at the n_1–n_2 interface that produces reflected wave 1 is a hard reflection resulting in a phase change. The transmitted part of the wave does not undergo a phase change at the n_1–n_2 interface, but it does under the hard reflection at the n_2–n_3 interface. There is no further phase change as it leaves the medium as a transmitted wave. Therefore, both paths have one phase change, and the phase change due to the extra distance travelled ($2t$) must be some number of half wavelengths for destructive interference, so the answer is (b). Note that the extra distance travelled is in medium 2, where the effective wavelength is λ/n_2.

21. a. The wave that eventually becomes 4 has a hard reflection at the n_2–n_3 interface, producing a phase change. There is no phase change at the other reflection, which is soft, or when it is transmitted into medium n_3. The wave that becomes 3 is a transmitted wave at both interfaces, so undergoes no phase changes at the interfaces. To have constructive interference between waves 3 and 4, the extra distance must correspond to a half number of wavelengths, so the correct answer is (b). Note that the extra distance travelled is in medium 2, where the effective wavelength is λ/n_2.

23. c. The maximum intensity will occur when the phase difference between the two waves is $\Delta\phi = 0$ radians. For the resultant intensity we have

$$I = I_1 + I_2 + 2\sqrt{I_1 I_2}\cos(\Delta\phi) = I_0 + 5I_0 + 2\sqrt{I_0 5I_0}\cos 0 = 6I_0 + 2\sqrt{5}I_0 = (6 + 2\sqrt{5})I_0$$

25. c. For a double-slit apparatus, we know that $d\sin\theta = m\lambda$. If we make d smaller, then $\sin\theta$ will get bigger, since m and λ do not change. The intensity of the light does not affect the location of the maxima and minima.

27. We first convert the frequency to Hz, then use $c = f\lambda$, and finally convert the wavelength to the desired units of nm:

469 THz = 4.69×10^{14} Hz

$$\lambda = \frac{c}{f} = \frac{3.00 \times 10^8 \text{ m/s}}{4.69 \times 10^{14} \text{ s}^{-1}} = 6.40 \times 10^{-7} \text{ m} = 640. \text{ nm}$$

This corresponds to visible light, and the colour for this wavelength is red.

29. Geometric optics can be used in cases where the electromagnetic radiation wavelength, λ, is much smaller than the size of the object to be investigated. The smallest dimension in this case is 1 mm.

    (a) $\lambda = 532$ nm $= 0.532 \times 10^{-3}$ mm $\ll 1$ mm

    Geometric optics can be used.

    (b) $\lambda = 5 \; \mu$m $= 5 \times 10^{-3}$ mm $\ll 1$ mm

    Geometric optics can be used.

    (c) $\lambda = 180$ nm $= 0.18 \times 10^{-3}$ mm $\ll 1$ mm

    Geometric optics can be used.

    (d) $\lambda = \dfrac{c}{f} = \dfrac{3.00 \times 10^8 \text{ m/s}}{88.0 \times 10^6 \text{ Hz}} = 3.41$ m $\gg 1$ mm

    Geometric optics cannot be used.

31. We will call I_0 the intensity of one wave alone. We calculate the intensity of the combined wave as follows:

    $$I = I_1 + I_2 + 2\sqrt{I_1 I_2} \cos(\Delta\phi) = I_0 + I_0 + 2\sqrt{I_0 I_0} \cos\left(\dfrac{\pi}{6}\right) = 2I_0 + 2I_0 \dfrac{\sqrt{3}}{2} = 3.73 I_0$$

    Therefore, the intensity of the combined wave will be about 3.7 times that of one wave alone.

33. The minimum result will be when the phase difference is π, while the maximum will be when the phase difference is zero. Calculating the intensity for the combined wave for each of these possibilities, we have

    $$I = I_1 + I_2 + 2\sqrt{I_1 I_2} \cos(\Delta\phi)$$
    $$I_{min} = I_0 + 9I_0 + 2\sqrt{9I_0^2} \cos\pi = 10I_0 - (2 \times 3I_0) = 4I_0$$
    $$I_{max} = I_0 + 9I_0 + 2\sqrt{9I_0^2} \cos 0 = 10I_0 + (2 \times 3I_0) = 16I_0$$

    Therefore, the combined intensity can range from $4I_0$ to $16I_0$.

35. The constructive interference condition for the Michelson interferometer is
    $m\lambda = 2d$

    In this equation, λ denotes the wavelength of the light, d is the distance the mirror moved, and m is an integer related to the number of bright fringes observed in the interference pattern. Hence, the minimum distance, d, the mirror moved is given by

    $$d = \dfrac{m\lambda}{2} = \dfrac{90 \times 633 \text{ nm}}{2} = 28485 \text{ nm} = 28.5 \; \mu\text{m}$$

37. We solve for the angle for the two extreme wavelengths (400 nm and 700 nm), converting all distances to metres:

$$\sin\theta = \frac{m\lambda}{d} \Rightarrow \theta = \sin^{-1}\left(\frac{m\lambda}{d}\right)$$

For $\lambda = 400$ nm:

$$\theta_{400} = \sin^{-1}\left(\frac{1 \times 4.00 \times 10^{-7}\,\text{m}}{4.50 \times 10^{-4}\,\text{m}}\right) = 8.89 \times 10^{-4}\,\text{rad}$$

For $\lambda = 700$ nm:

$$\theta_{700} = \sin^{-1}\left(\frac{1 \times 7.00 \times 10^{-7}\,\text{m}}{4.50 \times 10^{-4}\,\text{m}}\right) = 1.56 \times 10^{-3}\,\text{rad}$$

We can use trigonometry to calculate the interference distances on the screen. In the following, D is the distance to the screen, or 24.5 cm:

$$x_{400} = D\tan\theta_{400} = (24.5\,\text{cm})\tan(8.89 \times 10^{-4}\,\text{rad}) = 0.0218\,\text{cm}$$

$$x_{700} = D\tan\theta_{700} = (24.5\,\text{cm})\tan(1.56 \times 10^{-3}\,\text{rad}) = 0.0382\,\text{cm}$$

39. (a) We first use trigonometry to find the angle to the $m = 4$ maximum:

$$\tan\theta = \frac{2.40\,\text{cm}}{18.0\,\text{cm}}$$

$$\theta = \tan^{-1}\left(\frac{2.40\,\text{cm}}{18.0\,\text{cm}}\right) = 0.132\,55\,\text{rad}$$

Next we solve for the wavelength, remembering that this is an $m = 4$ maximum:

$$\lambda = \frac{d\sin\theta}{m} = \frac{(1.50 \times 10^{-4}\,\text{m})\sin(0.132\,55\,\text{rad})}{4} = 4.956 \times 10^{-6}\,\text{m} = 4960\,\text{nm}$$

(b) This wavelength is in the infrared region (see the electromagnetic spectrum in Chapter 27).

41. (a) As suggested in the hint, we can find the answer by setting up a spreadsheet and trying different slit spacings until the desired answer for the spacing is achieved. Alternatively, we can make an analytical approximation. We do this first below. Note that this is only valid if the angles are relatively small. We can calculate the angle separation, $\theta_2 - \theta_1$, between the $m = 1$ and $m = 2$ images in a double-slit experiment using the constructive interference condition, coupled with the small-angle trigonometric approximation:

$$\theta_2 - \theta_1 = \sin^{-1}\left(\frac{2\lambda}{d}\right) - \sin^{-1}\left(\frac{\lambda}{d}\right) \approx \frac{2\lambda}{d} - \frac{\lambda}{d} \approx \frac{\lambda}{d} \tag{1}$$

Using the result of (1), the spatial separation, d_{12}, between the $m=1$ and $m=2$ images is given by

$$d_{12} = D\tan\theta_2 - D\tan\theta_1 \approx D(\theta_2 - \theta_1) \approx \frac{D\lambda}{d} \quad (2)$$

From (2) it follows that

$$d \approx \frac{D\lambda}{d_{12}} = \frac{(0.750 \text{ m})(532\times10^{-9} \text{ m})}{4.00\times10^{-3} \text{ m}} = 9.975\times10^{-5} \text{ m} = 99.8\, \mu\text{m}$$

(b) Using (2), the spacing, d_{12}, for red light illumination is equal to

$$d_{12} \approx \frac{D\lambda}{d} \approx \frac{(0.750 \text{ m})(633\times10^{-9} \text{ m})}{9.975\times10^{-5} \text{ m}} = 4.76\times10^{-3} \text{ m} = 4.76 \text{ mm}$$

Alternative solution:

(a) It is possible to solve this problem without invoking the approximation represented in (1), instead using numerical methods. Let x be the unknown distance to the $m=1$ image on the screen relative to the zero-order image. Applying the double-slit condition for $m=1$, we have the following relationship (we omit units until the end for ease of calculation):

$$\sin\theta_1 = \frac{\lambda}{d} = \frac{x}{\sqrt{x^2 + 0.750^2}}$$

and, similarly, for the $m=2$ case:

$$\sin\theta_2 = \frac{2\lambda}{d} = \frac{x+0.004}{\sqrt{(x+0.004)^2 + 0.750^2}}$$

If we divide the first relationship by the second, and square each side, we obtain the following:

$$\frac{1}{4} = \frac{x^2[(x+0.004)^2 + 0.750^2]}{(x+0.004)^2(x^2 + 0.750^2)}$$

We can use a spreadsheet (or computer program or computational application) to try different values of x to make the two sides equal. When we do this, we find that $x = 0.00400$. Using this value for x gives $\theta_1 = 0.3056°$:

$$d = \frac{\lambda}{\sin\theta_1} = \frac{532\times10^{-9} \text{ m}}{\sin(0.3056°)} = 9.97\times10^{-5}\text{ m} \text{ (in agreement with the other method, within rounding)}$$

(b) $\sin\theta_1 = \dfrac{1\lambda}{d}$

$\theta_1 = \sin^{-1}\left(\dfrac{633\times 10^{-9}\text{ m}}{9.97\times 10^{-5}\text{ m}}\right) = 0.3638°$

$\sin\theta_2 = \dfrac{2\lambda}{d}$

$\theta_2 = \sin^{-1}\left(\dfrac{2(633\times 10^{-9}\text{ m})}{9.97\times 10^{-5}\text{ m}}\right) = 0.7276°$

$x_1 = 0.750\tan\theta_1 = 0.750\tan(0.3638°) = 0.004762\text{ m}$
$x_2 = 0.750\tan\theta_2 = 0.750\tan(0.7276°) = 0.009525\text{ m}$

Therefore, the difference in x values is 0.004763 m, which considering significant digits of the difference should be rounded to 0.0048 m. This is slightly different from the approximate method. In the answers we have used these slightly more accurate values.

43. We first find the grating spacing from the inverse of the number of lines per length and then use $c = f\lambda$ to find the wavelength from the frequency. Then we are ready to use the diffraction grating relationship to find the angle:

$d = \dfrac{1}{N} = \dfrac{1}{5000.\text{ lines/cm}} = 2.000\times 10^{-4}\text{ cm} = 2.000\times 10^{-6}\text{ m}$

$f = 500.\text{ THz} = 5.00\times 10^{14}\text{ Hz}$

$\lambda = \dfrac{c}{f} = \dfrac{3.00\times 10^8\text{ m/s}}{5.00\times 10^{14}\text{ s}^{-1}} = 6.00\times 10^{-7}\text{ m}$

$\sin\theta = \dfrac{m\lambda}{d}$

$\theta = \sin^{-1}\left(\dfrac{1\times 6.00\times 10^{-7}\text{ m}}{2.000\times 10^{-6}\text{ m}}\right) = 17.5°$

45. The grating spacing is simply the inverse of the number of lines per length:

$d = \dfrac{1}{N} = \dfrac{1}{3000.\text{ lines/cm}} = 3.333\times 10^{-4}\text{ cm} = 3.333\times 10^{-6}\text{ m}$

We then calculate the angles for the two wavelengths:

$\sin\theta_{486} = \dfrac{m\lambda}{d}$

$\theta_{486} = \sin^{-1}\left(\dfrac{1\times 4.86\times 10^{-7}\text{ m}}{3.333\times 10^{-6}\text{ m}}\right) = 0.146\text{ rad}$

$$\sin\theta_{656} = \frac{m\lambda}{d}$$

$$\theta_{656} = \sin^{-1}\left(\frac{1\times 6.56\times 10^{-7}\text{ m}}{3.333\times 10^{-6}\text{ m}}\right) = 0.198 \text{ rad}$$

If we call the distance to the screen $D = 15.0$ cm, we can use trigonometry to find the distance, x, from the zero-order image to the first-order image:

$$x = D\tan\theta$$
$$x_{486} = 15.0 \text{ cm} \times \tan(0.146 \text{ rad}) = 2.21 \text{ cm}$$
$$x_{656} = 15.0 \text{ cm} \times \tan(0.198 \text{ rad}) = 3.01 \text{ cm}$$
$$\Delta x = 3.01 \text{ cm} - 2.21 \text{ cm} = 0.80 \text{ cm}$$

47. We denote the angles corresponding to the first-order bright interference fringe for the two extreme wavelengths $\lambda_1 = 360$ nm and $\lambda_2 = 675$ nm by θ_1 and θ_2, respectively. Similarly, we denote the corresponding distances from the centre to the first-order interference fringe by d_1 and d_2. From the geometry of the spectrometer characterized by the distance, $D = 5.00$ cm, between the detector and the diffraction grating, we can write the following relationships:

$$\sin\theta_1 = \frac{d_1}{\sqrt{D^2 + d_1^2}} \tag{1}$$

$$\sin\theta_2 = \frac{d_2}{\sqrt{D^2 + d_2^2}} \tag{2}$$

Also,

$$d_2 - d_1 = 2.25 \text{ cm} \tag{3}$$

The constructive first-order interference conditions for the diffraction grating characterized by N lines per unit length and wavelengths λ_1 and λ_2 are

$$\sin\theta_1 = N\lambda_1 \tag{4}$$

$$\sin\theta_2 = N\lambda_2 \tag{5}$$

Squaring the ratio of (1) and (2), and then the ratio of (4) and (5), we obtain

$$\left(\frac{d_1}{d_2}\right)^2 = \frac{D^2 + d_1^2}{D^2 + d_2^2}\left(\frac{\lambda_1}{\lambda_2}\right)^2 \tag{6}$$

Denoting distance d_1 by x, *keeping all distances in centimetres*, and using (3), we can write (6) as

$$\frac{x^2}{(2.25+x)^2} = \frac{25.0+x^2}{25.0+(2.25+x)^2}\left(\frac{360}{675}\right)^2 = \frac{25.0+x^2}{25.0+(2.25+x)^2}\left(\frac{64}{225}\right) \qquad (7)$$

After some simplification, (7) is equivalent to

$$225x^2[25+(2.25+x)^2] = 64(25+x^2)(2.25+x)^2$$
$$2576x^4 + 11\,592x^3 + 77\,441x^2 - 115\,200x - 129\,600 = 0 \qquad (8)$$

Using software, we find the positive solution of (8):

$$x = d_1 = 1.774 \text{ cm} \qquad (9)$$

Using (1) and (4) with this value, it follows that

$$N = \frac{d_1/\lambda_1}{\sqrt{D^2+d_1^2}} = \frac{1.774\times10^{-2}\text{ m}/360\times10^{-9}\text{ m}}{\sqrt{(5.00\times10^{-2}\text{ m})^2+(1.774\times10^{-2}\text{ m})^2}} = 928\,800\text{ m}^{-1} = 929\text{ mm}^{-1}$$

49. We find the wavelength in the medium where the extra distance is travelled, so $n = 1.30$. Making reference to the figure, we have placed circles at the places where phase-changing hard reflections take place. Each path has one (π radian) phase change, so the *relative* phase constant for the two paths is zero, and we use the top row of Table 29-1.

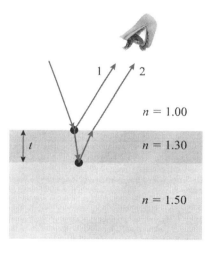

We are seeking a solution for constructive interference, so we use the left-hand column:

$$2t = m\frac{\lambda}{n} \quad m = 1, 2, 3, \ldots$$

We want the three smallest thickness values, so we use $m = 1$, $m = 2$, and $m = 3$. We will leave the wavelength in nanometres, meaning the thickness will also be expressed in nanometres:

$$t = \frac{m\lambda}{2n}$$

for $m = 1$: $t = \frac{1 \times 589 \text{ nm}}{2 \times 1.30} = 227$ nm

for $m = 2$: $t = \frac{2 \times 589 \text{ nm}}{2 \times 1.30} = 453$ nm

for $m = 3$: $t = \frac{3 \times 589 \text{ nm}}{2 \times 1.30} = 680.$ nm

51. (a) Each path has one hard reflection, so both have a π rad phase shift due to reflections (see diagram below with hard reflections marked with a circle).

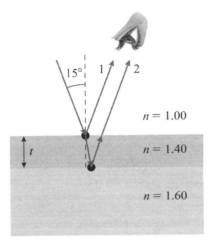

Since both have the same phase shift, there is a phase constant difference of zero and we use the first row of Table 29-1. We are interested in destructive interference and therefore use the right-hand column of Table 29-1. It is the wavelength in the medium where the extra distance is travelled that we use, so that is where the index of refraction is $n = 1.40$:

$$2t = \left(m + \frac{1}{2}\right)\frac{\lambda}{n} \quad m = 0, 1, 2, 3, \ldots$$

We want the minimum thickness, so we use $m = 0$. If we leave the wavelength in units of nanometres, then the thickness will also have these units:

$$t = \frac{\lambda}{4n} = \frac{550 \text{ nm}}{4 \times 1.40} = 98 \text{ nm}$$

(b) If we don't make the perpendicular incidence assumption, then the situation becomes

$$t = \frac{\lambda \cos(15°)}{4n} = \frac{550 \text{ nm} \times 0.966}{4 \times 1.40} = 95 \text{ nm}$$

As expected we need a bit thinner film now when we take into account that the slightly inclined path that increases the distance travelled.

53. (a) Assuming a hard reflection on both interfaces by the antireflection coating, we can calculate the minimum coating thickness, t, using the destructive interference condition:

$$t = \frac{\lambda}{4n} = \frac{1250 \times 10^{-9} \text{ m}}{4 \times 1.38} = 226 \text{ nm}$$

(b) The constructive interference condition is

$$2t = m\frac{\lambda}{n} \Rightarrow \lambda = \frac{2nt}{m}$$

For the first order of the constructive interference ($m = 1$), the above equation gives

$$\lambda = \frac{2 \times 1.38 \times 226 \text{ nm}}{1} = 624 \text{ nm}$$

55. We can calculate the width, w, of the slit using the destructive interference condition for a single-slit diffraction:

$$w \sin\theta = m\lambda \Rightarrow w = \frac{m\lambda}{\sin\theta}$$

$$w = \frac{1 \times (589 \times 10^{-9} \text{ m})}{\sin(1.15°)} = 2.93 \times 10^{-5} \text{ m} = 29.3 \, \mu\text{m}$$

57. First we will find the angle to the first ($m = 1$) minimum. We must put the slit width and the wavelength into the same units, which will be metres for our solution:

$$\sin\theta = \frac{m\lambda}{w}$$

$$\theta = \sin^{-1}\left(\frac{1 \times 5.89 \times 10^{-7} \text{ m}}{4.50 \times 10^{-4} \text{ m}}\right) = 1.31 \times 10^{-3} \text{ rad}$$

Now we will use trigonometry to find the distance, x, from the centre to the minimum on one side. The distance from the slit to the screen is D:

$$\tan\theta = \frac{x}{D}$$

$$x = D\tan\theta = 50.0 \text{ cm} \times \tan(1.31 \times 10^{-3} \text{ rad}) = 6.55 \times 10^{-2} \text{ cm}$$

The entire width of the image is $2x$:

image width $= 2x = 0.131$ cm

59. We count only bright fringes within the central maximum corresponding to the single-slit diffraction pattern. The first-order destructive interference condition for single-slit diffraction gives

$$\sin \theta = \frac{\lambda}{w} \qquad (1)$$

The constructive interference condition for the double-slit experiment gives

$$\sin \theta = \frac{m\lambda}{d} \qquad (2)$$

From (1) and (2), it follows that

$$\frac{\lambda}{w} = \frac{m\lambda}{d} \Rightarrow m = \frac{d}{w} = \frac{0.125 \text{ mm}}{0.0150 \text{ mm}} = 8.33$$

Eight bright fringes can be seen on each side. Therefore, including the central peak, $2 \times 8 + 1 = 17$ bright fringes are visible.

61. Using the Rayleigh criterion, the smallest angle θ_r separating two points that can be resolved with light of wavelength λ and using a circular aperture of diameter D is

$$\sin \theta_r = 1.22 \frac{\lambda}{D}$$

Assuming that the average light wavelength the eye is most sensitive to is $\lambda = 550.0$ nm and $D = 2.85$ mm, then

$$\theta_r = \sin^{-1}\left(1.22 \times \frac{550.0 \times 10^{-9} \text{ m}}{2.85 \times 10^{-3} \text{ m}}\right) = 0.0135°$$

If the distance between the observer and the car is ℓ and the distance between the lights is $d = 1.45$ m, the following geometrical relationship is valid:

$$\tan\left(\frac{\theta_r}{2}\right) = \frac{d/2}{\ell} \Rightarrow \ell = \frac{d}{2\tan\left(\frac{\theta_r}{2}\right)} = \frac{1.45 \text{ m}}{2 \times \tan(0.0135°/2)} = 6.15 \times 10^3 \text{ m} = 6.15 \text{ km}$$

63. The Rayleigh criterion for the minimum resolvable angle for a circular aperture is

$$\sin \theta_r = 1.22 \frac{\lambda}{D}$$

Therefore, the angle is larger, of poorer resolution, for a longer wavelength, and we will use the largest wavelength (650 nm) in doing this problem. First we calculate the diameter of the lens using the focal length and the focal ratio:

$$f_{\text{ratio}} = \frac{f}{D}$$

$$D = \frac{f}{f_{\text{ratio}}} = \frac{4.15 \times 10^{-3} \text{ m}}{2.20} = 1.89 \times 10^{-3} \text{ m}$$

$$\sin \theta_r = 1.22 \frac{\lambda}{D}$$

$$\theta_r = \sin^{-1}\left(1.22 \frac{6.50 \times 10^{-7} \text{ m}}{1.89 \times 10^{-3} \text{ m}}\right) = 4.196 \times 10^{-4} \text{ rad} = 2.40 \times 10^{-2\circ}$$

65. (a) It varies from the orange–red end of the visible spectrum (0.600 μm) to the infrared.

    (b) Applying the Rayleigh criterion, we have for the limiting resolution angle

    $$\sin \theta_r = 1.22 \frac{\lambda}{D}$$

    $$\theta_r = \sin^{-1}\left(1.22 \frac{28.5 \times 10^{-6} \text{ m}}{6.50 \text{ m}}\right) = 5.35 \times 10^{-6} \text{ rad} = 3.06 \times 10^{-4\circ}$$

    $$= 3.06 \times 10^{-4\circ} \times \frac{3600''}{1°} = 1.10''$$

    (c) Since the angle is so small, we can readily use the definition of radian in finding the dimension of the smallest resolvable object:

    $$x \approx 5.35 \times 10^{-6} \text{ rad} \times 120 \text{ ly} \times \frac{9.46 \times 10^{15} \text{ m}}{1 \text{ ly}} \times \frac{1 \text{ AU}}{1.50 \times 10^{11} \text{ m}} = 40.5 \text{ AU}$$

67. The bright fringes counted when going from evacuated to containing the gas are due to the phase difference between the two arms of the Michelson interferometer that is induced by the wavelength-reducing shift (λ/n) due to the light passing twice through the arm containing the gas with index of refraction n. If the length of each arm is denoted by ℓ, this means that the equivalent length of the arm filled with gas is equal to $n\ell$. Given that the only phase shift is due to the presence of the gas, the constructive interference condition is
$m\lambda = 2\ell(n-1)$

    It follows that

    $$n = 1 + \frac{m\lambda}{2\ell} = 1 + \frac{205 \times 633.000 \times 10^{-9} \text{ m}}{2 \times 0.365000 \text{ m}} = 1.000\,18$$

69. The constructive interference condition for the double-slit experiment using light of wavelength λ and distance d between the two slits is
$d \sin \theta = m\lambda$

Based on this equation, the angular separation, θ_1, between the $m=0$ and $m=1$ spots is given by

$$\theta_1 = \sin^{-1}\left(\frac{\lambda}{d}\right) = \sin^{-1}\left(\frac{264 \times 10^{-9} \text{ m}}{0.100 \times 10^{-3} \text{ m}}\right) = 0.151°$$

71. If the angle between one $m=2$ maximum and the centre is denoted by θ_2, then the angle between the two $m=2$ maxima is equal to $2\theta_2 = 1.30°$. Hence, $\theta_2 = 0.650°$. We can calculate the spacing between the slits, d, from the constructive interference condition:

$$d = \frac{2\lambda}{\sin\theta_2} = \frac{2(480 \times 10^{-9} \text{ m})}{\sin(0.650°)} = 8.46 \times 10^{-5} \text{ m} = 85 \text{ }\mu\text{m}$$

73. (a) Let θ_1 and θ_2 denote the deflection angles for the first-order interference fringes corresponding to the wavelengths $\lambda_1 = 350$ nm and $\lambda_2 = 1000$ nm, respectively. Using the constructive interference condition for the diffraction grating, we can calculate angles θ_1 and θ_2 as follows. We first calculate the slit spacing, d, from the number, N, of lines per unit length:

$$d = \frac{1}{N} = \frac{1}{600\,000} \text{ m} = 1.667 \times 10^{-6} \text{ m}$$

$$\theta_1 = \sin^{-1}\left(\frac{\lambda_1}{d}\right) = \sin^{-1}\left(\frac{3.50 \times 10^{-7} \text{ m}}{1.667 \times 10^{-6} \text{ m}}\right) = 12.1°$$

$$\theta_2 = \sin^{-1}\left(\frac{\lambda_2}{d}\right) = \sin^{-1}\left(\frac{1.000 \times 10^{-6} \text{ m}}{1.667 \times 10^{-6} \text{ m}}\right) = 36.9°$$

(b) The effective width of the detector, w, is equal to

$$w = 2048 \text{ pixels} \times 14 \frac{\mu\text{m}}{\text{pixel}} = 28\,672 \text{ }\mu\text{m} = 28.7 \text{ mm}$$

(c) Given the detector width, w, calculated in part (b) and the distance, D, between the grating and the screen, the following geometric relationship is valid:

$$w = D(\tan\theta_2 - \tan\theta_1)$$

Hence, the distance, D, in millimetres, is equal to

$$D = \frac{w}{\tan\theta_2 - \tan\theta_1} = \frac{28.7 \text{ mm}}{\tan(36.9°) - \tan(12.1°)} = 53.5 \text{ mm}$$

75. The constructive interference condition for the reflected light rays between the two glass plates separated by distance t is given by

$$2t = (2k+1)\frac{\lambda}{2}, \ k = 0,1,2,\ldots \quad (1)$$

In (1), a hard reflection of light on the glass plate equivalent to a 180° phase shift was assumed, since there is a hard reflection at air to glass, but not glass to air. Equation (1) can also be written as

$$t = \left(m - \frac{1}{2}\right)\frac{\lambda}{2}, \ m = 1,2,3,\ldots \quad (2)$$

The maximum distance between the glass plates is $t_{max} = 75.0 \ \mu m$. Hence, using (2), the maximum number of bright fringes that can be observed is given by

$$m = \frac{1}{2} + \frac{2t_{max}}{\lambda} = \frac{1}{2} + \frac{2 \times 75.0 \times 10^{-6} \ m}{532 \times 10^{-9} \ m} = 282.5 \quad (3)$$

We truncate 282.5 to a whole number, 282, of fringes.

77. (a) The Rayleigh criterion for a circular aperture of diameter D and light wavelength λ gives the angular resolution θ_r as

$$\theta_r = \sin^{-1}\left(1.22\frac{\lambda}{D}\right)$$

Assuming $\lambda = 550.0$ nm, the value of θ_r for the Canada-France-Hawaii Telescope is

$$\theta_r = \sin^{-1}\left(1.22 \times \frac{550.0 \times 10^{-9} \ m}{3.60 \ m}\right) = 0.000\,010\,68° = 0.000\,0107°$$

(b) $\theta_r = 0.000\,0106\,8° = 0.000\,0106\,8° \times 3600\frac{1''}{1°} = 0.0384''$

79. The array approximates a telescope with a much larger effective diameter, D. In this case, the diameter can be approximated as being equal to the distance between the two telescopes: $D \approx 3000$ km. The angular resolution, θ_r, given by the Rayleigh criterion is

$$\theta_r = \sin^{-1}\left(\frac{\lambda}{D}\right) = \sin^{-1}\left(\frac{c}{Df}\right) = \sin^{-1}\left(\frac{3.00 \times 10^8 \ m/s}{3.00 \times 10^6 \ m \times 1.42 \times 10^9 \ Hz}\right) = 4.03 \times 10^{-6}°$$

Note that in this case it is not a circular aperture.

81. We first express the limiting resolution in degrees:

$$\theta_r = 1.5'' \times \frac{1°}{3600''} = 4.167 \times 10^{-4}°$$

The Rayleigh criterion relating the angular resolution, θ_r, for a circular aperture of diameter D with radiation wavelength λ is

$$\sin\theta_r = 1.22\frac{\lambda}{D}$$

It follows that

$$D = \frac{1.22\lambda}{\sin\theta_r} = \frac{1.22 \times 500 \times 10^{-9}\text{ m}}{\sin(4.167 \times 10^{-4\circ})} = 0.0839\text{ m} = 8.4\text{ cm}$$

Based on the precision of the 1.5″, the final answer should be expressed to two significant digits, if we assume that the 500 nm assumed wavelength was intended to be precise to at least that much.

83. The destructive interference condition for the diffraction on a slit of width w is

$$w\sin\theta = m\lambda \quad (1)$$

Therefore, the angle, θ_1, of deflection corresponding to the first dark fringe ($m=1$) is given by

$$\theta_1 = \sin^{-1}\left(\frac{\lambda}{w}\right) \quad (2)$$

We can now specify the condition that the first three bright fringes originated from the double-slit interference pattern within angle θ_1 on each of the two sides away from the centre:

$$d\sin\theta_1 = 3\lambda \quad (3)$$

Using (2) and (3), we obtain

$$d\frac{\lambda}{w} = 3\lambda \Rightarrow w = \frac{d}{3} = \frac{0.125\text{ mm}}{3} = 41.7\ \mu\text{m}$$

Note that the wavelength cancels out and does not matter for the final result.

85. Values for the refractive index of kerosene vary, but it is at least slightly higher than that of water. A value of 1.39 is most frequently given for kerosene, with the index of water about 1.33. Therefore, the thin film reflection situation is as shown on the next page.

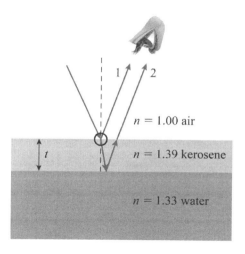

We have a hard reflection at the air–kerosene interface but a soft reflection at the kerosene–water interface. This means that the two paths have a phase constant difference of π rad (not counting the phase difference due to the extra distance travelled). In Table 29-1, we therefore use the bottom row. We want constructive interference, so we use the left column, with the following condition:

$$2t = \left(m + \frac{1}{2}\right)\frac{\lambda}{n} \quad m = 0,1,2,3,\ldots$$

We will assume the thinnest film, $m = 0$, and therefore have the following for the thickness. Note that the extra distance travelled is in the kerosene, so $n = 1.39$:

$$t = \frac{\lambda}{4n} = \frac{500 \text{ nm}}{4 \times 1.39} = 89.9 \text{ nm} = 8.99 \times 10^{-8} \text{ m}$$

The question does not specify, but we will reasonably assume the spill is circular of radius r. We can use trigonometry to compute r from the angle subtended (see figure below):

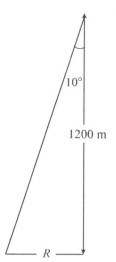

$$\tan(10°) = \frac{R}{1200 \text{ m}}$$
$$R = 1200 \text{ m} \times \tan(10°) = 212 \text{ m}$$

From this we can calculate the surface area of the spill:

$$A = \pi R^2 = \pi (212 \text{ m})^2 = 1.41 \times 10^5 \text{ m}^2$$

If we multiply this by the thickness of the spill, we get the volume of kerosene:

$$V = \pi R^2 t = \pi (212 \text{ m})^2 \times 8.99 \times 10^{-8} \text{ m} = 0.0127 \text{ m}^3$$

The 1200 m. is expressed to two significant digits. If we assume that the "approximately 500 nm" wavelength is intended to also be precise to two significant digits, then we would express the final answer to two significant digits as 13 L. Given the assumptions in the problem, it could reasonably be argued that only one significant digit is warranted, in which case the final answer would be expressed as about 10 L. We assumed that it was the thinnest possible film producing constructive interference ($m = 0$). If we had assumed $m = 1$, the result would be three times as much. We also assumed a circular shape for the spill.